# the night sky

# the night sky

a guide
to observing
the Sun, Moon
and planets

Steve Massey

First published in Australia in 2003 by
New Holland Publishers (Australia) Pty Ltd
Sydney • Auckland • London • Cape Town

14 Aquatic Drive Frenchs Forest NSW 2086 Australia
218 Lake Road Northcote Auckland New Zealand
86 Edgware Road London W2 2EA United Kingdom
80 McKenzie Street Cape Town 8001 South Africa



National Library of Australia Cataloguing-in-Publication Data:
Massey, Steve.
The Night Sky: a guide to observing the Sun, Moon and planets.

Bibliography.
Includes index.
ISBN 1 74110 083 6.

1. Astronomy. 2. Astronomical instruments. I. Title.

522

Publishing Manager: Robynne Millward
Project Editor: Karen Gee
Designer: Karlman Roper
Production Controller: Jane Kirby
Printer: Leefung-Asco, China

10 9 8 7 6 5 4 3 2 1

# About the author

Steve Massey is an associate member of the Astronomical Society of Australia and has been a dedicated amateur astronomer for some thirty years, his main interest being in the study of the Solar System. Steve worked at the Anglo-Australian Telescope at Siding Spring Observatory some years ago and frequently returns to the facility, where he continues to explore the cost-effective possibilities of sharper, ground-based imaging with video.

Steve's widely published pictures of Mars in 2001 were considered among some of the best ever taken from a ground-based telescope of its class, and his wonderful images of the planets have been published around the globe.

Initially published in the United States, Steve was the primary co-author of the first-ever book on the topic of video astronomy and has been a contributor to several other publications worldwide, including an astronomy encyclopedia and a historical guide to the Moon. He has also contributed to a number of general observing guides and has several magazine articles to his credit.

An occasional speaker at schools, universities and on radio, Steve has developed interactive learning software tools about the Moon, and other astronomy-related applications.

# Contents

The Lagoon Nebula (M8) in Sagittarius is a star-birth hothouse which is visible with the naked eye. It is nebulae like this where new solar systems evolve within dark and cool collapsing clouds. This simple backyard photograph was taken with a standard 35mm SLR camera and Kodak 1000 Gold film at the prime focus of an f/4.5 250mm (10in) reflector telescope.

# Preface

There is nothing quite like the peace and tranquillity of a dark night sky filled with stars and the sounds of crickets. To lie on the ground in outback Australia and gaze upward at midnight is an experience you'll never forget. What is all this that surrounds us? What does it all mean? Where do I fit in to it all? These are questions most of us whisper to ourselves at such reflective moments.

Since the time of Galileo our species has become increasingly aware of our own almost insignificant and transient existence within the realm of this immense universe, a universe still veiled in so much mystery. But we ourselves are an incredible miracle, spawned from the building blocks of stars and blessed with a consciousness to look analytically skyward and question all things that abound. When gazing into the eyepiece of a telescope we are among the privileged few on this planet to experience our own personal view into the depths of space. It offers an ineffable sense of being out there among the stars.

Turning my first humble telescope to the Moon in 1971 marked a turning point in my life. Utterly mesmerised by the striking details of craters, mountain ranges and dark wrinkled plains across its surface, I soon found myself drawing what I saw in the eyepiece. But perhaps the most inspiring personal experiences were my first views of the planets. Among the myriad of stars, which revealed nothing more than an infinite point of light, were a handful of star-like objects that were truly distinguishable as other worlds suspended in the void of space.

Way before my first glimpses through a telescope, humankind was already well entrenched in the space age. *Pioneer 10* was approaching Jupiter around the time I first witnessed that planet's tiny disc through a small 30X telescope. Over the years the United States and Russia would send an armada of planetary probes into space equipped with scientific instruments and cameras, shattering old myths and revealing much of the true nature of our planetary family. Long-held mysteries, theories and hopes for life elsewhere in the Solar System were radically changed forever as a plethora of data and new, glossy, high-resolution pictures were being radioed back from our space-faring robotic reporters. In this new millennium we are treated almost daily to marvellous new pictures, from photographs of the planets, taken from modern spacecraft, to views of distant galaxies at the edge of known space and time through the eyes of the Hubble Space Telescope.

Over the years I have met people who believed that to view celestial wonders as seen in glossy publications requires huge telescopes like those straddling the mountain tops in Hawaii and Siding Spring in north-western New South Wales. But when they first see a fuzzy patch in the sky resolved into innumerable stars or the rings of Saturn through a common backyard telescope, sighs of almost disbelief often follow. On the other hand, I have also met others who embarked on the hobby with the purchase of a low-cost telescope and unrealistic expectations of what they might see. This is largely the result of misleading advertising and a lack of personal endeavour to understand at least the basics of when and how to locate objects in the sky and how those objects will appear in a small telescope.

Topically, and in terms of research, astronomy has grown significantly over the years, leading to many diverse pursuits for astronomers both professional and amateur. In Australia and overseas, many patrons of the telescope seek out faint, distant deep-sky targets either for their observational challenge or simply as photographic trophies. Others enjoy the challenges of observing dynamically changing targets much closer to home. An active international space program (led mostly by the United States) since the 1960s has left few mysteries for the amateur to ponder, but unlike the static appearance of distant stars, galaxies and nebulae, the Solar System is alive with activity and much of it can be observed with a telescope from your own backyard. With dedicated personal endeavour there is always the possibility of witnessing and reporting a new development as-yet unseen by others. This might be a new comet, asteroid or dust storm on Mars!

While spacecraft must travel for many months or years to reach the planets, we telescopic travellers can view them all within minutes by simply sweeping our humble light-gathering instruments across the night sky and from the comfort of our own backyards. To observe the planets we don't need to travel great distances seeking dark-sky locations; most of them are so bright we can even observe them through thin cloud.

From our personal observing sites and with the aid of special filters, we can witness huge sunspots and solar flares so large they could engulf the earth. With almost any small telescope we can indulge in magnificent nightly views of the Moon and its vastly diverse terrain. We can follow the progress of eruptions high in the atmosphere of Jupiter or witness the striking transits of its moons. We can watch the rotation of Mars and gaze upon its icy polar caps and hazy clouds or maybe witness the progress of a huge global dust storm. A beginner's telescope easily reveals distant Saturn and its magnificent rings, almost 1.5 billion kilometres away. Even Titan, Saturn's largest moon, can be seen changing position nightly. You can follow the apparitions of Venus and Mercury, the inner worlds, as they move through phases like the Moon and occasionally cross the face of the Sun in rare transit events. You can even observe the planets during the day. There is so much to see and do that you'll soon find yourself absorbed for hours.

Today, amateur astronomers are capable of producing magnificent photographs of the planets and beyond which rival many of the best pictures taken from some of the world's largest telescopes many years ago. Home computers, software and digital imaging cameras have made this possible. Even with the common household camcorder or digital still camera we can record wonderful pictures of the Sun, Moon and planets. In addition to these new and improved tools for the amateur, affordable commercial telescopes today are far superior both optically and mechanically to many of those of the past. With this in mind, however, one must still be aware of optical performance limits. Poorly manufactured beginner's telescopes in years past resulted in a waning pursuit of the hobby by many adventurous newcomers. Unstable mounts and poorly crafted optics made simple observing an unenjoyable effort. But competitive manufacturing today has changed much of this, raising the overall quality of lower-priced telescopes. Despite criticisms of the beginner's telescopes you may inevitably encounter within the amateur astronomy fraternity, most of those I have reviewed in recent years produce terrific images, but only within a realistic magnification range. Everyone has a budgetary limitation and not everyone has the dexterity to build their own 'near perfect' telescope; therefore we must each of us learn to get the most out of what we have.

Amateur astronomy is comprised of many interesting individuals from all walks of life, ranging from airline pilots, doctors, lawyers, mechanics and storekeepers to guys who moonlight in rock-and-roll bands. The hobby is only as limited as the challenges we set ourselves. Whether your interest is in casual stargazing or more serious pursuits, this book is your tour guide to our celestial neighbourhood. But be warned—planetary astronomy is not only fun, interesting and rewarding, it can become an addiction!

Steve Massey

# Acknowledgments

Compiling your thoughts and experiences in the form of a book is always a challenge and every effort has been undertaken to ensure I have passed on to the reader an accurate introductory guide to this fascinating hobby. I am very grateful to friends and colleagues for their input and comments on the manuscript. Any inaccuracies that may be evident are my responsibility alone.

I would like to thank those who have contributed in various ways to achieving this general guide. They include the director and support team of the Research School of Astronomy and Astrophysics (RSAA), also known as Mount Stromlo Siding Spring Observatories (MSSSO), for granting me telescope time over the years. It was deeply saddening to see the destruction of valuable technology, resources and of course the wonderful historic telescopes of Mount Stromlo during the devastating Canberra bushfires of January 2003.

Special thanks to Steve Quirk, an accomplished Australian astrophotographer and an enthusiastic colleague who provided valued comments and suggestions on the manuscript, along with various pictures and astrophotography exposure tables used in the book. My sincere thanks to the team at Tasco Australia and especially to Kevin Johnson, Ian James and Rob McIntyre for providing me with telescopes used in much of the performance and observational evaluations and comparisons for the purposes of this book.

To those friends and family who have supported my interest, answered questions, supplied material, published my work or simply lent equipment for testing, my gratitude and thanks to: Vince Ford (RSAA), Jonathon Nally, Richard Newton, Robert McNaught, Rick Fienberg and Gary Seronik (*Sky & Telescope* magazine), Graeme Stewart, Julie Houghton, Richard Fahy, Michael Fairbrass, Pat Watson and Logan Shield. Special thanks also to Scott Massey for assisting with and providing the excellent diagrams throughout.

In today's world it has become more parentally difficult to guide children in the pursuit of constructive or educational after-hours activities. To this end, I would also like to acknowledge my parents, Bob and Shirley, for supporting and encouraging my hobby through the early years.

Putting a book together requires a dedicated publisher and publishing team. I extend my deepest gratitude to New Holland Publishers Australia for making this book come to life. In particular my thanks and gratitude to general manager Fiona Schultz for adopting the project. A big thanks also to my editor Karen Gee for her excellent editorial detective work and suggestions while ironing out the wrinkles in the manuscript, and to senior designer Karl Roper for developing the finished product and for illustrations advice. Thanks also to Lesley Pagett and Kate Richards for their warm and enthusiastic support.

And finally, to my incredibly supportive wife Sandra, who has demonstrated time and again what it truly means to love someone and share a life together. Since embarking on my first book in 1999 and several articles that followed, she has continued to support my late-night keyboard tapping and long nights at the eyepiece. To you ... my love and eternal thanks.

# How to use this book

This book serves as an introductory guide for anyone wanting to learn more about the wonders of astronomy and stargazing, and will lead you, step by step, through the Solar System and its family of planets.

First, you'll find a concise history of planetary astronomy, followed by a guide to understanding how and where the stars and planets are placed and can be found in the sky. Next we examine the two most prominent celestial objects, the Sun and the Moon. The Sun is dealt with first since it is central to all other bodies and is indeed the largest object in the Solar System. Next we take an observationally rich tour of the Moon, which is often the first port of call for most beginners.

Chapters for the planets have been organised by order of their respective orbits, or distance, from the Sun, so that we can examine each along an outward journey of the Solar System. For the beginner the easiest planetary subjects to observe, in terms of their apparent size in a telescope, are Jupiter and Venus (at favorable times), followed by Saturn and Mars. These chapters are great starting points before setting your sights on more challenging targets such as Mercury, the extreme distant worlds of Uranus, Neptune and Pluto, and the asteroids. With this in mind, some chapters are more content-rich than others where greater historical or observational discussion may be useful. For example, features of the Moon and Mars are dealt with in more detail since there is so much more to be gleaned from the eyepiece than for, say, Uranus, Neptune or Pluto.

If you are seeking to better understand how your telescope functions, want to find out which accessories you may need, or would even like to venture into the exciting pursuit of astrophotography, then you will find sections on these at the rear of the book under 'Observing tools' and 'Creating planetary portraits'.

# Part one: In the beginning

Before exploring the individual aspects of our closest celestial neighbours, it is interesting to learn the early perceptions and thinking about the mechanics of the stars and planets as seen through the eyes of our ancestors. In the following chapters we'll briefly look at the progress of early astronomers and their attempts to understand and explain various phenomena. We'll review our current understanding of the Solar System and its evolution, then you'll find information on how the sky is mapped and navigated so you can find the planets, their moons and other celestial bodies among the myriad of stars. You'll even learn how to observe some of the planets during the daytime.

# Battling with the spheres

There's little doubt that at sometime or another, whether returning home from an outing or simply putting the garbage out after dusk, we have marvelled at the beauty of a pristine, clear night sky. Since you are reading this book it is most likely you are a conscious, or even subconscious, skyward scanner seeking to know a little more about those wandering and sometimes very bright points of light that pass overhead each night.

Most people simply refer to these sparkling points of light as the stars. Any description beyond the Moon, stars and even falling stars may hold little appeal for die-hard romantics, but for the inquisitive there is much more than meets the eye. Although the majority of these points of light overhead are truly stars, a few among them are not. To understand a little of how we have come to discriminate these heavenly family members from all the other stars we need go back in time to when the ancient science of astronomy was practised in its most basic form.

To many cultures in ancient times, the fixed star patterns of the night sky represented characters of mythology. However, these beacons of the night also promised many practical benefits. Used by travellers for navigational purposes, the monthly westward progression of the stars also proved to be a reliable annual calendar of the seasons. Among the stars, however, were five that did not seem to hold fixed positions as the others did. These five moved in mysteriously independent ways. Two of these stayed very near to the Sun, never seeming to trace a path all the way across the night sky. Only visible after sunset or before sunrise, these two mystical travellers would certainly add confusion when it came to deciphering the motions of the planets. The other three moved eastward from month to month with varying speed among the fixed stars. One of these 'stars' presented the most obvious irregular motion, appearing to stop for a spell then move backwards for a brief time, only to stop yet again and continue along its original eastward path. What mysterious magic made this one travel with such non-conforming motion among the other stars? These nomads of the night sky were called 'planets', an early Greek name for 'wanderers'.

In ancient and medieval times the most modern thinkers within the 'developing' world emerged from many countries and cultures of the Northern Hemisphere. Philosophers, mathematicians and other practitioners of basic sciences began to grow in places such as Asia, the Middle East and other northern provinces. What we know today regarding the mechanics behind the wandering motions of our

Under dark country skies, the Milky Way (the galaxy in which we live) traces a hazy path across the sky and the number of visible stars seems countless. Beyond that which our eyes can see, even more stars are revealed. In this simple 10-minute exposure, looking towards the centre of our galaxy, several nebulae can also be seen. These floating clouds of gas and dust are stellar nurseries for new stars, planets and moons.

Sun, Moon and the planets against the background stars is attributed to a number of lateral-thinking individuals from these times, each building upon the earlier work of others. Many have been memorialised with prominent craters on the Moon named after them.

Before the telescope, we could only observe our environment within the limits of our eyes and from the heights of the tallest mountains. From our earthly perspective it is not hard to understand how easily we were misled into believing that Earth was stationary and all other heavenly bodies circled around it. A number of complex systems and theories were devised in order to account for the positions and different motions of the Sun, Moon, planets and stars in the scenario of an Earth-centred universe. Aristotle (384–322 BC), a student of the great Greek philosopher Plato (427–348 BC), proposed a system of nested, transparent crystal spheres encompassing a motionless, spherical Earth at its centre. The Moon moved

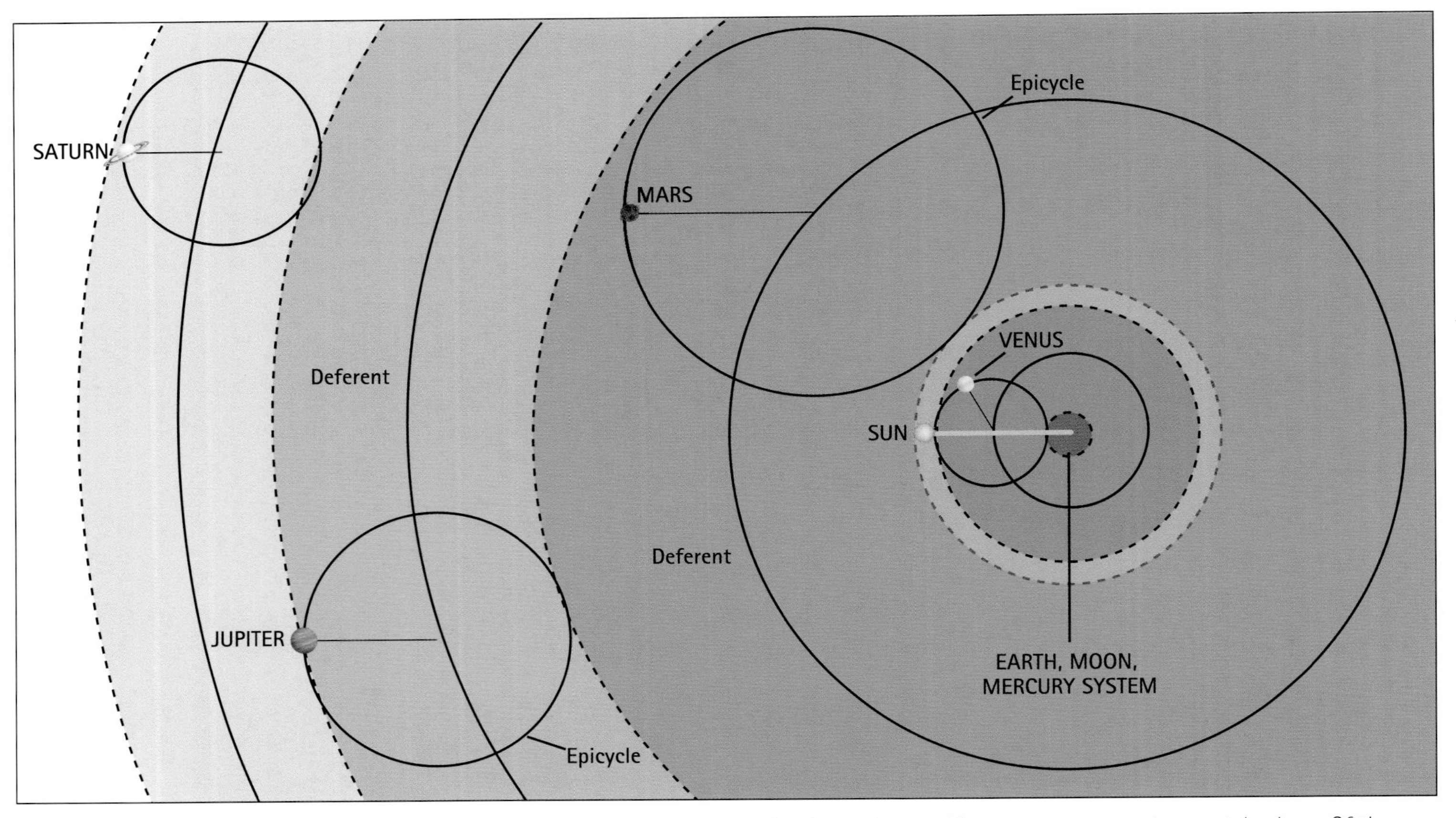


Early beliefs placed the Earth at the centre of the visible universe while the fixed stars were said to rotate on an outer crystal sphere. Of the variations that would follow, this basic diagram of the Ptolemaic system shows how early astronomers accounted for the wandering motions of the planets using epicycles and deferents with uniform movement. Nearer to Earth, the Moon and Mercury respectively were also depicted in this fashion.

about the Earth upon the innermost sphere while the other spheres, of ever-increasing diameter, represented the Sun, Mercury, Venus, Mars, Jupiter, Saturn and finally the fixed background stars, in that order. While this concept was welcomed with open arms for many years thereafter, it did not account for the retrograde (backwards) motions of the planets and was later modified by Ptolemy.

Ptolemy (127–151 AD) devised a system of large concentric circles (*deferents*), still with Earth at its centre. The outward order of celestial orbits then followed with the Moon, Mercury, Venus, Sun, Mars, Jupiter and Saturn. The retrograde motions of the Moon and planets were accounted for by placing smaller circles (*epicycles*) along their respective orbital paths. This awkward system would become the standard for years to come, but although there were many other variations to follow, none could produce accurate predictions without some need to jig the system. There had to be a better solution and one that could account for all the anomalies.

## Turning point

After hundreds of years of staggeringly slow advances in astronomy, the truth about the heliocentric (sun-centred) nature of our Solar System began to emerge around the 15th and 16th centuries. Even though a heliocentric system had indeed been hypothesised by Aristarchus of Samos (310–230 BC), it is interesting to ponder how far we might have evolved today, both scientifically and culturally, had our ancestors readily adopted his suggestion so long ago.

These were exciting though perhaps controversial times, still largely controlled by religious overlords. It was certainly the dawning of a new era for astronomy. The most notable contributions at this time were those of Copernicus, Tycho Brahe (known as Tycho), Kepler and Galileo. Nicholas Copernicus (1473–1543) produced the controversial publication *De Revolutionibus* which defined the heliocentric system we know today, and placed the known planets in their appropriate orbits about the Sun.

Johannes Kepler (1571–1630) was a brilliant mathematician of his time who worked diligently to solve the irregularities in observed planetary behaviour over the purely circular orbits of the Copernican system. This required reliable observational data and so he utilised the meticulous records of Tycho Brahe (1546–1601), who was renowned for his accurate measurement of the positions of the planets and stars. In doing so he developed three key physical laws of planetary motion, known as Kepler's Laws, thus uncovering the conundrum of

Galileo Galilei employed one of the first known telescopes in a quest for greater understanding of the celestial spheres. With this exciting new magnifying tool he observed the changing phases of Venus and discovered that Jupiter had orbiting moons of its own. These discoveries fuelled his convictions in support of a Sun-centred system over the long-held belief that everything revolved around the Earth.

what we now know to be elliptical orbits of varying individual characteristics.

Italian-born Galileo Galilei (1564–1642) was a student of medicine before turning his interests to astronomy in later years. Throughout historical literature he is probably the most renowned figure in popular terms, although perhaps no more deserving than Copernicus, Tycho or Kepler. Galileo's convictions in support of a heliocentric system were driven by his advocacy for the mathematical works of Copernicus. Galileo employed one of the first-known telescopes to reveal both the imperfect nature of the Moon's relatively rough terrain and a Sun with patchy black spots. Among other observational discoveries, Galileo first reported the existence of moons orbiting a planet other than Earth in 1610. These were the four largest moons of Jupiter. Originally named the Medicean stars by Galileo, in recognition of his financial sponsors of the time, they later became known as the Galilean moons. They may also be referred to as the Jovian moons. Furthermore, his observations of lunar-like phases during different apparitions of Venus lent further credence to the Copernican system. Historians suggest that Galileo was quite a character, who would openly challenge and even ridicule ignorant authorities within the church. His discoveries served a crushing blow to Aristotelians and due to his rebellious, mocking style he made several enemies. Prohibited by the Catholic Church from teaching Copernicanism as truth in 1616, Galileo later defied this prohibition, which ultimately led to his condemnation and house arrest in 1633. He was confined until his death some years later. Although his works were later allowed to be published it took the church almost 360 years before they would publicly make a formal apology. In 1992 Galileo was finally exonerated and recognised for his groundbreaking contributions to astronomy.

# The new era

With the development of the telescope in the early 1600s the doors to new knowledge and further discoveries were rapidly opening. Great improvements in optical performance soon followed with larger light-gathering instruments being built, mostly funded by the wealthy. Using a variety of refracting or reflecting telescopes, many distinguished names in early planetary studies were soon to take their place in history, including Christiaan Huygens (1629–1693), Jean Domenique Cassini (1625–1712), Edmond Halley (1656–1742), William Herschel (1738–1822), Giovanni Schiaparelli (1835–1910) and Percival Lowell (1855–1916).

The planets were no longer points of starlight but small discs revealing subtle, sometimes vivid markings. The rings of Saturn were soon unveiled along with a number of its moons. New moons around Jupiter were being found along with more detailed views of its encircling cloud belts and a giant, cyclone-like storm known as the Great Red Spot. Observers were now witnessing the polar ice caps of Mars, the red planet, along with strange dark surface features hinting at the possibility of life beyond Earth. In 1781 a new planet, Uranus, would also be discovered, effectively doubling the size of the known Solar System from around 1.3 billion kilometres to around 2.8 billion kilometres. And in 1846 yet another planet, Neptune, was to be discovered far beyond the orbit of Uranus. Neptune's largest moon, Triton, was detected only three weeks after the discovery of the planet itself. It wasn't until 1930 that tiny Pluto was discovered. The most distant of all the known planets, frigid Pluto orbits the Sun once every 249 years. Due to its somewhat diminutive size compared with all the other major planets, its status as a 'true' planet remains a controversial topic for planetary scientists.

# Collapsing cloud

After centuries of analytical skyward scanning, one major question still begged an answer: how did this all come about? How indeed was the Solar System created? A number of theories have been mooted over the years. However, modern physics, mathematics, state-of-the-art observational tools and space probes continue to gather more convincing evidence to support the now widely accepted evolutionary process called the accretion theory.

Within the realm of our known universe are clusters of enormous galaxies that float like islands in the vast barrenness of space and time. They each gravitationally

In the sword of the constellation of Orion, the Great Orion Nebula, visible to the naked eye, is a colourful stellar nursery of hydrogen and oxygen gases and dust, spawning new stars and planets.

interact with one another. Our own galaxy, the Milky Way, is one of a number of other galaxies in a cluster we call the Local Group. The Milky Way is a spiralling entity of gas and dust clouds that are the incubators for all the stars and planets we can see and those we cannot, hidden from our view forming within dark, nebulous cocoons. The accretion theory proposes that our Solar System condensed from one of these irregular patches of cloud drifting through the void of cold interstellar space in one of the great spiral arms of our galaxy. The self-destruction of a nearby star perhaps sent shock waves reeling into the cold primordial dust cloud, forming massive globules or knots of dark, rich material. Over time and under the mounting forces of gravity, this irregular cloud of material eventually collapsed into a relatively flat disc—the accretion disc. Slowly swirling like a whirlpool of water, most of the nebulous material was drawn into the dimly glowing bulge at its centre where pressures were mounting and a new star (our Sun) was forming. In the outer regions of the disc, molecules of gas and dust began to clump together, growing even larger through the natural chaotic processes of gravitational attraction.

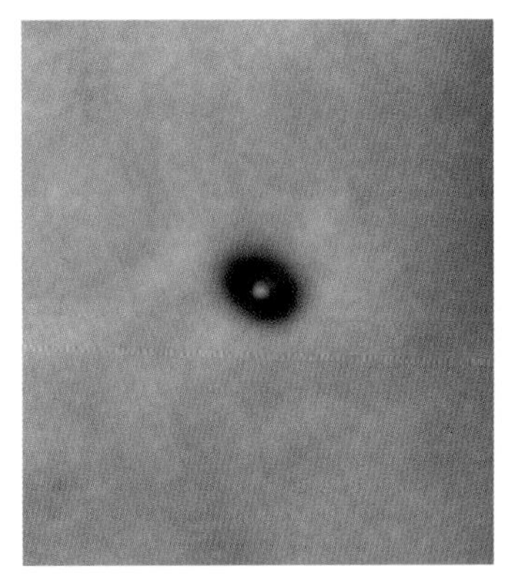

This image from the Hubble Space Telescope reveals one of several protoplanetary discs discovered forming in the Orion Nebula almost 1500 light years away. The central red glow is a young, newly formed star. *(Courtesy M McCaughrean, CR O'Dell and NASA)*

A dust particle became a grain of sand that would later evolve to the size of a rock. Rocks collided; some attached themselves to others, growing further in size. These rocky clusters became mountain-sized planetesimals and the mountains eventually became worlds.

Heavier elements, such as iron, and rock formed near the inner Solar System while the lighter gaseous and icy materials condensed as the major components of the outer planets. Each newly forming world moulded itself into an imperfect rocky, icy or gaseous sphere spinning on its own axis. As the glowing red nucleus at the centre of this newly forming solar system became denser it also grew hotter, eventually triggering a nuclear fusion reaction that would give birth to our Sun. At this time all the remaining rubble of less-than-substantial size was blown outward, as if by a huge gust of wind, into the Solar System. Some of this rocky debris collided with the newly forming planets; almighty impacts left deep scars in the form of basins or craters.

Some of this rubble continues to wander aimlessly throughout the Solar System—increased gravitational perturbations and destiny will determine their final violent encounters. Between the orbits of Mars and Jupiter a portion of this rubble has settled into a region known as the asteroid belt. Perhaps these are the remains of a small forming world ripped apart by the enormous gravitational influence of a nearby encounter with Jupiter. The planets grew in size as they swept their way around the Sun, eventually cleaning up most of the remaining rubble—craters lie testament to the final rocky bombardment of the inner planets and most of the solid mud-ice worlds beyond.

The Solar System is comprised of nine major planets, seen here in order of their respective orbital distances from the Sun. The minor rocky planets of the asteroid belt reside between the orbits of Mars and Jupiter. (diagram not to scale)

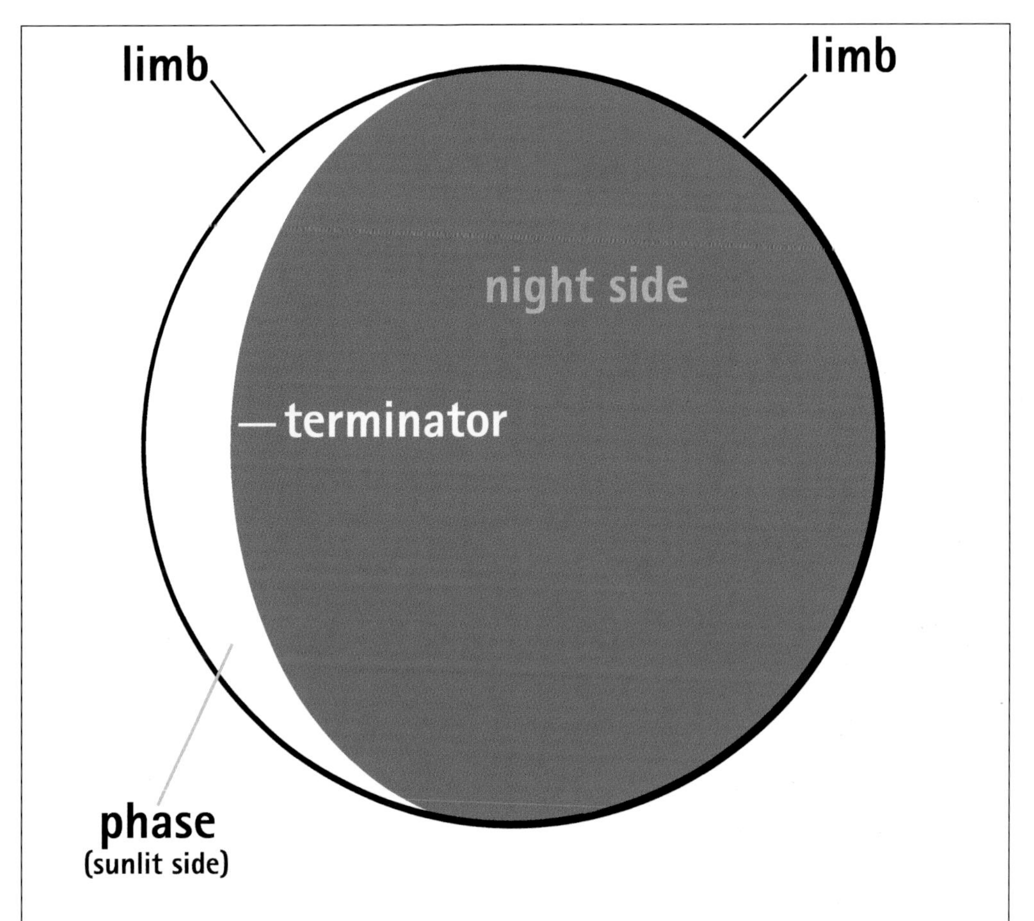

As the Sun illuminates an inner planet or the Moon, as seen from Earth, the reflected light waxes and wanes through various phases. The limb of a planet or the Moon is the outer edge of its disc while the terminator is the boundary between night and day.

By the end of this dramatic stage show, occurring some 5 billion years ago, the nine planets we know today became fixed in their orbits about the Sun, orbits which move in the same direction as the Sun's axial rotation. On an outward journey from the Sun we encounter the orbits of 'Moon-like' Mercury then cloud-covered Venus; our beautiful blue world, the Earth; the red planet, Mars; massive Jupiter; the ringed world of Saturn; mysterious Uranus; Neptune, the 'other' blue planet; and, furthermost, icy Pluto. Whether technically dead or alive, each world is uniquely different. Born from the same star stuff, they are testament to the laws of natural physics and the incredible beauty that is creation.

# The great illuminator

Our Sun is the great illuminator, shining like a fiery beacon at the centre of the Solar System. The light it emits travels through the vacuum of space at 299 792 kilometres per second. The planets and their moons reflect its light with varying intensities. This reflective property is referred to as an object's *albedo*. It is based on a scale of 0 to 1, 0 being the darkest and 1 a perfect reflecting surface. This scale is sometimes quantified in terms of a percentage.

From our earthbound perspective, the positions of the planets and the Moon relative to the Earth and Sun dictate how much illuminated surface can be seen. To understand this, try to imagine zooming upward into space from the North Pole, then looking down at the Solar System. Earth is the third planet from the Sun and from this new perspective we also see two inner orbiting planets, Mercury and Venus, quite differently from the six outer planets. Through a telescope Mercury and Venus can be seen to change from fully illuminated discs to a crescent phase like our Moon. The outer planets are always seen as full or near-full discs but never half-illuminated or crescent.

To understand how this works you can suspend a white ping-pong ball from a length of string to represent a planet. Now turn the lights off and sit yourself at eye level to a lit candle on a table. The candle is our Sun and you are Earth. Position the ping-pong ball directly opposite you on the far side of the candle; be careful not to burn your hand! Note how the ping-pong ball appears fully illuminated. Now move it slowly anti-clockwise around the candle. Note how its illuminated side now starts to shrink like the phases of the Moon. As the ball passes directly between you and the candle, the side facing you is now completely shaded from the candle's light. As you continue moving the ball, you will note the opposite side gradually catching the light from the candle. This is how we see the inner planets Mercury and Venus because they move around the Sun along orbits inside that of the Earth.

Now let's simulate an outer planet like Mars or Jupiter. Facing away from the candle hold the suspended ping-pong ball directly in front you. You many need to hold it slightly above head height to avoid your shadow interrupting. Note that it appears fully illuminated by the candlelight. If possible, move yourself in a circle around the candle, keeping your eye on the ball, so to speak. Note that the ball remains fully illuminated. Of course, the outer planets are not always directly opposite the Earth and can present some slight amount of change in phase, but never more than a sliver of darkness on either side. You can demonstrate this by staying put and, while continuing to look at the ball, manoeuvring it around you and your candle Sun. Note that when it is situated at right angles to you on either side of the candle it shows slightly less than a fully illuminated globe.

To follow the changing phases of the Moon, place the ping-pong ball directly between you and the candle so that it appears completely shaded. This simulates a new moon. Now move the ball clockwise and note how it moves through waxing phases just like the Moon. Once you are facing away from the candle, the ball appears completely illuminated like that of a full moon (so long as your head isn't in the way). Continue turning and the phases appear to wane until we eventually reach our new moon position once more.

The outer edge of the Moon or a planet's disc is referred to as the *limb*, while the boundary that divides the illuminated and shaded sides of the disc or sphere is called the *terminator*.

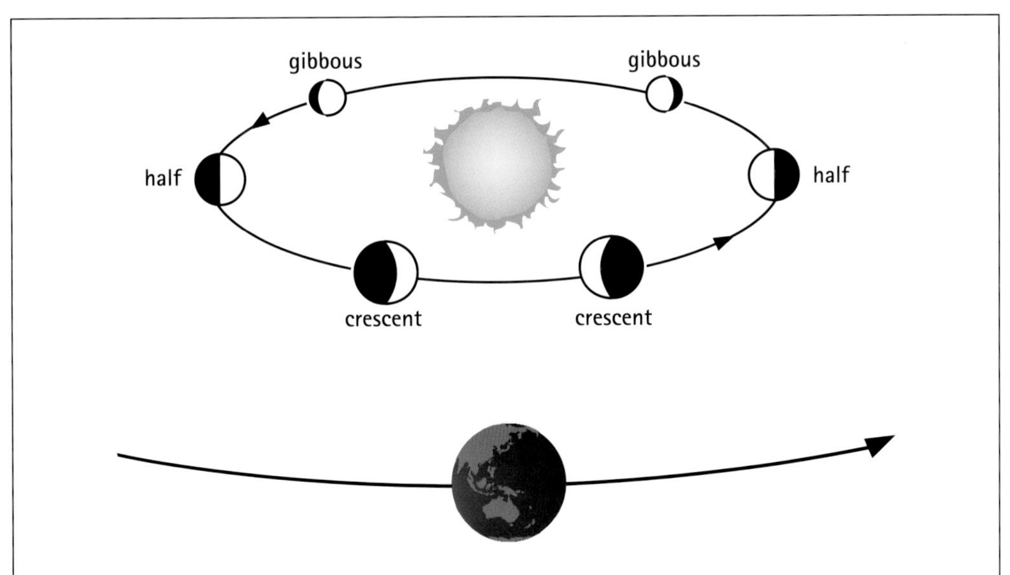


As seen from Earth, the inner orbiting planets Mercury and Venus present changing phases like those of the Moon. The outer planets (not shown in this diagram) present only a gibbous or near fully illuminated disc, but never a half or crescent phase like the inner worlds. Of the outer planets, Mars exhibits the greatest phase effect.

# Navigating the night sky

Coming back down to Earth, so to speak, things take on a somewhat different perspective. It is important to understand a little about the sky overhead and how the stars are placed in it so we can make sense of the constantly changing positions of the Moon and planets among them.

## The celestial sphere

From any location on the Earth, the sky overhead appears like a huge dome. During the day the Sun appears to ride around this great curve in the sky, while at night the stars appear as though fixed to a giant black sphere revolving around us from east to west. This imaginary sphere is actually a projection of the Earth in the sky and is called the *celestial sphere*. The stars and the Sun appear to move across

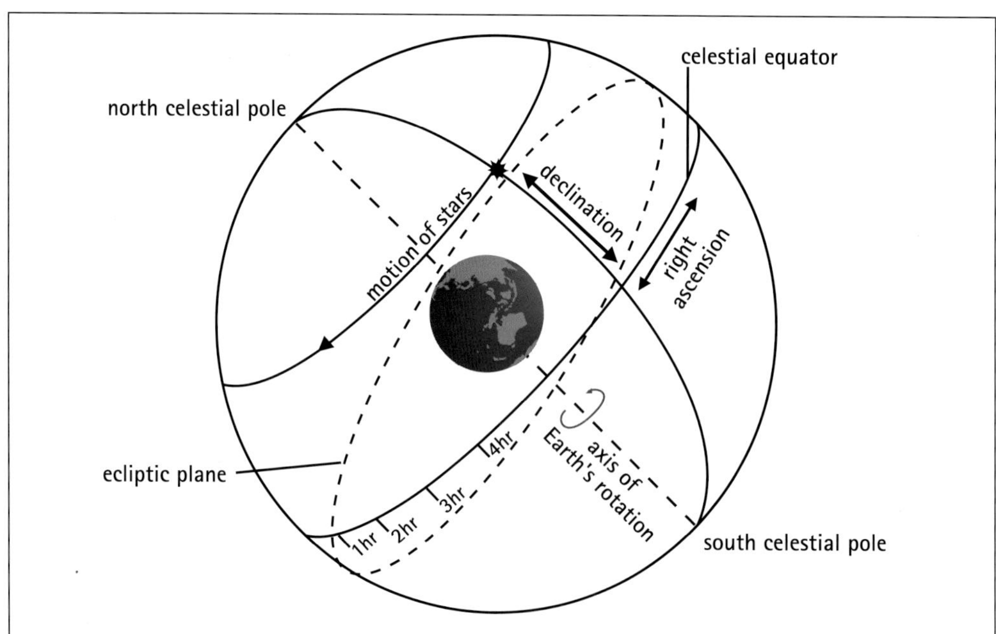


The celestial sphere. The celestial equator is a projection of the Earth's equator on the sky along with the ecliptic, which is the plane of Earth's orbit around the Sun. The planets lie close to the ecliptic plane.

it because the Earth is rotating. The Earth rotates around its poles, called the celestial north and south poles. The celestial poles are sometimes referred to as true north and south and are situated several degrees from magnetic north (slightly to the west) and magnetic south (slightly to the east). It is important to know where they are situated for correct polar alignment of your telescope. This is particularly essential for long-exposure astrophotography or CCD imaging (see page 190).

A 35mm SLR camera pointed towards the south celestial pole reveals the Earth's axial rotation in an exposure of little more than 5 minutes. An equatorial telescope must be polar aligned to the centre of this spinning wheel of stars to accurately track objects in the night sky.

In addition to the commonly understood north, south, east and west coordinates, a number of imaginary lines crossing the celestial sphere have been created for the purposes of navigating it.

If you look directly overhead, wherever you are standing, this point in the sky is called the *zenith*. It is at right angles or 90° to the horizon from all directions. The opposite end, or diametrically opposed point, to zenith (i.e. directly below zenith) is called *nadir*. Now imagine a great circle that curves across the sky from the celestial north pole, through zenith overhead, down to the celestial south pole then through the nadir point on the other side of the globe back up to

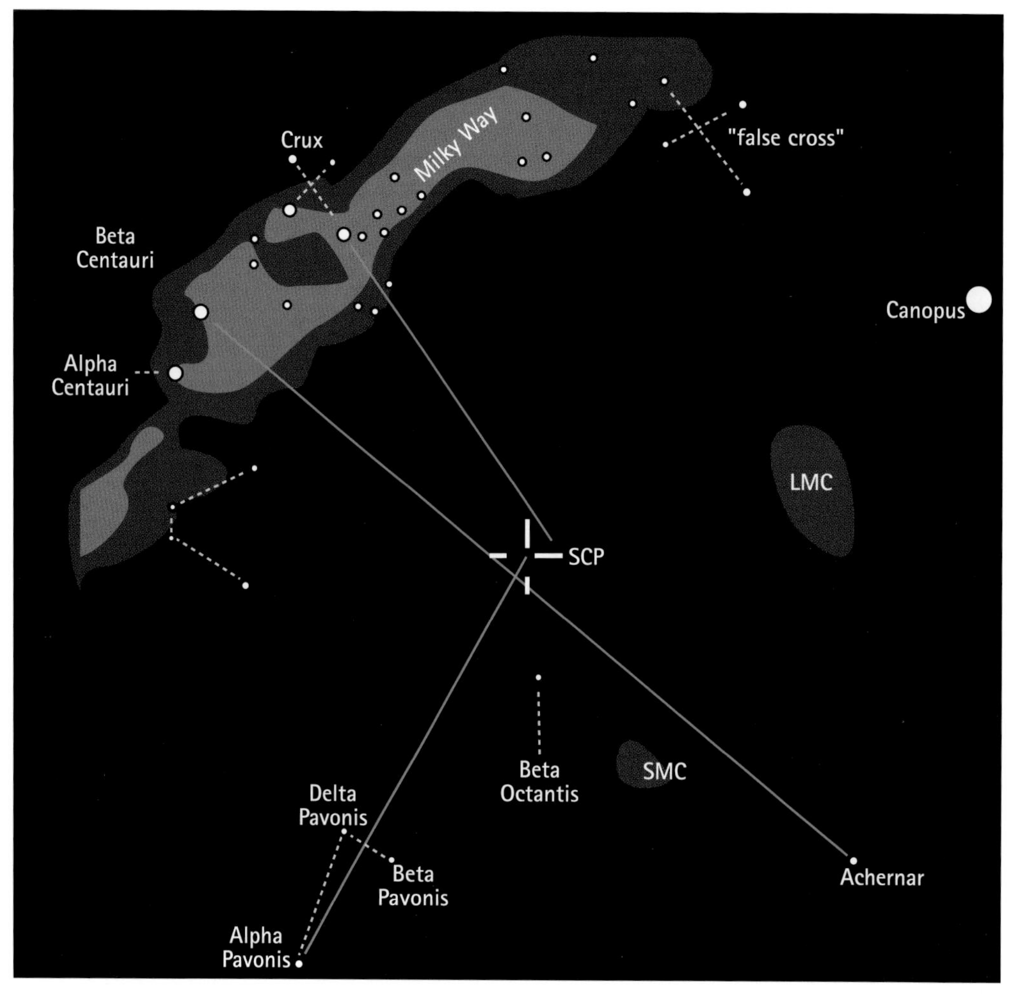


Correct polar alignment of an equatorial telescope is important for enjoyable protracted observing and photography. The elevation of the south celestial pole (SCP) in degrees above the horizon depends on the observer's latitude south of the equator. This diagram is a simple guide to rough alignment, which is generally satisfactory for most purposes. A mount with a polar guide scope for the stars of Octans will provide more accurate alignment.

celestial north. This is called the *meridian* and essentially divides the sky into Western and Eastern Hemispheres. Another imaginary line projected on to the celestial sphere is that of the Earth's equator, aptly called the *celestial equator*. This divides the Northern and Southern Hemispheres.

Perhaps through some catastrophic event in Earth's evolutionary past, our planet's axis of rotation is tilted by 23.5° to the plane of its orbit around the Sun.

It is due to this tilt that we experience opposite seasons in each hemisphere. Thus when it is summer and autumn in the Southern Hemisphere, it's winter and spring in the Northern Hemisphere.

During one Earth orbit (one year), the centre of the Sun crosses the celestial equator twice, at which point each hemisphere moves into either autumn or spring. On this important day, the length of night and day are, for a short time, roughly the same. The point at which this occurs is known as the *equinox*. When the Sun crosses the celestial equator from south to north this is called the vernal equinox and in the Southern Hemisphere we move into autumn. When the Sun crosses from north to south we move into spring. The entrenched term used by astronomers is the autumn equinox at this time because the name was originally conveived in the Northern Hemisphere, corresponding to its change of season, but for the Southern Hemisphere this is technically a spring equinox.

Another factor relating to the Earth's tilt affects how high the planets and Sun are seen in the sky. This is yet another imaginary line called the *plane of the ecliptic*, which we'll discuss later.

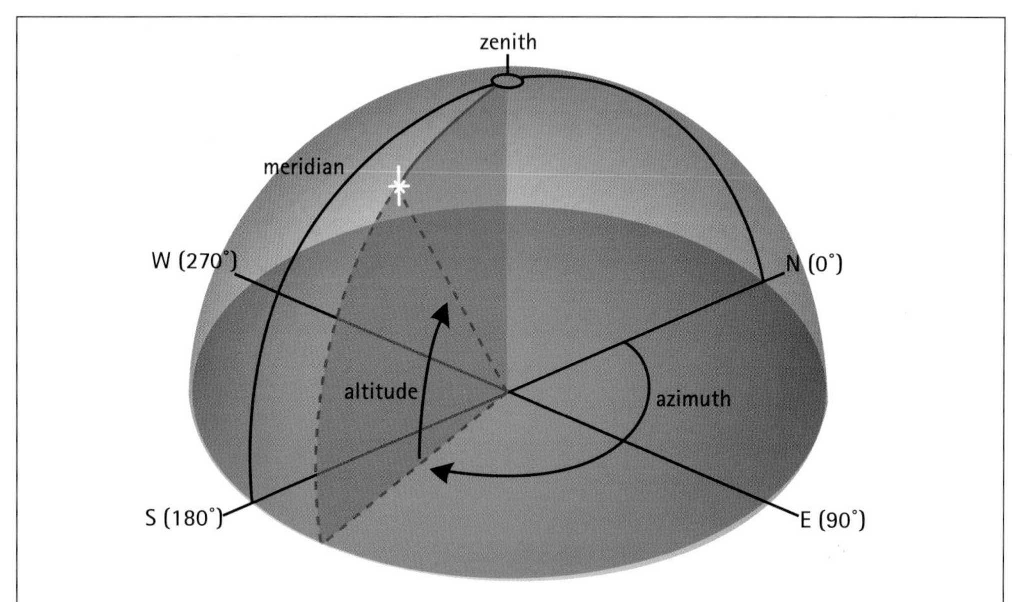


Altazimuth coordinates can be used to locate a celestial object at a given time and from a given location. 'Alt', or altitude, is given from 0° at the horizon to 90° overhead while 'azimuth' is the horizontal position measured through 360° right of designated 0° north. The meridian effectively divides the celestial dome of the sky into Eastern and Western Hemispheres while zenith is the point directly above the observer.

# Finding your way around the sky

As the science of astronomy evolved over time, systems for mapping and locating the positions of celestial objects were developed to aid both navigators and astronomers alike. The two methods discussed further on are based on an equatorial system and a horizon system, both of which depend on the observer's location and time. First we must know where we are on the globe and in what time zone we are situated.

## Longitude, latitude and time

Like a navigational meshing around the globe of the Earth, our planet it is subdivided by imaginary lines of longitude and latitude. The effect of longitude is directly related to time along Earth's west-to-east daily rotation. For the convenience of worldwide travellers and communications, it was essential that time zones be allocated to various regions across these lines of longitude around the globe. In 1884, Greenwich in the United Kingdom was selected as the *meridian* (0 hours) reference point for timekeeping and thus the time at Greenwich came to be known as Greenwich Mean Time (GMT). For scientific purposes, GMT is referred to as Universal Time (UT) and is the standard by which an observer records his or her observations. On the globe of earthly time, Sydney, Melbourne and Brisbane are situated 10 hours ahead of UT. When recording your observations from locations along this timeline, 10 hours must be deducted from the current local time to obtain UT date and time. Observers along the timeline for Perth will of course deduct 8 hours from local Perth time. For all participating regions, daylight saving must also be taken into account.

Latitude relates to your location in degrees north (+) or south (–), from 0° at the equator to 90° at the poles. Observers at the same latitudes but different longitudes will have the same view of the stars overhead and at the horizon but at different times. Your location in latitude has a fundamental effect on your view of the sky and also determines the height at which the north or south celestial poles are elevated. When polar aligning an equatorial mounted telescope, you must adjust the tilt of the mount in accordance with your observing latitude so that the optical tube runs parallel to the axis of the poles. To ascertain your location in longitude and latitude, you can look up a good world map or refer to a computer planetarium program and select from a list of major cities. For meticulously serious observers, a hand-held Global Positioning System (GPS) will provide the most accurate coordinates. Some manufacturers of GOTO telescopes (which allow users to locate targets from a computer-based hand controller at the press of a button) provide GPS systems built into the drive electronics.

## The equatorial system

Right ascension and declination coordinates tell us where a celestial body is on a fixed map of the sky.

*Right ascension* (RA) is the term given to the east-to-west motion of the celestial sphere. Since Earth completes one axial rotation through 360° every 24 hours, the sky can subsequently be divided into hours of arc. Hence one hour of arc represents 15°. For refined positional accuracy, each hour is further subdivided into minutes and seconds of arc. For example, 1° of sky is equal to 60 arc minutes and 1 minute of sky is equal to 60 arc seconds. The apparent disc size of the Sun and Moon equates to roughly 0.5° of sky or 30 arc minutes. Since the planets are visually so much smaller, their relative sizes are quoted in arc seconds. A single quote mark (′) is usually used to denote an arc minute. An arc second is denoted by a double quote mark (″).

*Declination* (dec) is the term given to the angular distances in degrees from the celestial equator (0°) to the celestial north pole (+90°) or celestial south pole (–90°).

A celestial object is designated coordinates for right ascension and declination where the two points intersect in the sky. For example, Alpha Crucis or Acrux (the bottom-most star of the Southern Cross) can be found at RA 12 hours, 27 minutes and declination 63°. Due to the sheer distances involved, stars appear deceptively stationary from our earthly perspective and are assigned to fixed coordinates for right ascension and declination. However, right ascensions and declinations for much nearer Solar System bodies are constantly changing.

Polar aligned telescopes on equatorial mounts can locate and track celestial targets in right ascension and declination. Correctly aligned to the Earth's celestial poles and with the polar angle axis adjusted for your observing latitude, this type of mount facilitates smooth tracking of objects across the sky as the Earth's diurnal motion (24-hour axial rotation) wheels through the celestial sphere around us. Objects may be kept in the field of view for protracted periods with the aid of a variable speed motorised drive. Another useful accessory is a variable speed declination drive for making minor tracking error adjustments and slewing in that axis. This is particularly essential for long-exposure astrophotography or CCD imaging.

## The horizon system

In the horizon system, the coordinates for a celestial target are determined for your local sky at that moment.

*Altazimuth* refers to the vertical coordinate 'alt' or 'altitude' from 0° at the horizon to 90° overhead, and 'azimuth' is the horizontal position measured through 360° right of designated 0° north. Telescopes with altazimuth mounts come in various forms, like the classic tripod fork mount, the Dobsonian swivel box-style platform or the computerised GOTO fork mount common to many Schmidt-Cassegrain telescopes. Owners of Dobsonian telescopes can retro-fit the mount with an automated electric motor drive system to alleviate the need for constant manual recentring of the telescope each time the subject drifts from the field of view. These systems are available in kit form. Some models offer computer interfacing, allowing a telescope to be controlled remotely from a laptop or desktop PC using various astronomy software programs which are often supplied.

# Finding the planets

So how do we know where to look for the planets? How can we recognise them among the stars? Unlike our early ancestors we don't need to rely purely on visual means to spot vagabond stars among those which appear fixed in their positions. If you're unfamiliar with the night sky there are many useful resources available today to help make sense of the patterns of the stars and planets. Indeed, the planets can confuse a newcomer when trying to identify a constellation from a chart of bright stars. Monthly and annual *ephemerides* (tables for positions of celestial objects) are published, providing day-by-day rise and set times and coordinates for locating the planets.

Monthly astronomy periodicals also provide handy star charts with positions for the planets, comets and asteroids plus upcoming meteor shower events. Today we also have the luxury of numerous astronomy software programs for computers. Several programs are available for download on the Internet as freeware or shareware. Most of these programs provide sufficient information to satisfy even the most advanced amateur astronomer. Laptop computers and modern palm pilots can also drive many of these programs, offering the portability of a humble star wheel with the power of modern computing. In addition to these great tools, GOTO telescopes also contain comprehensive databases. At the push of a button, the telescope will slew to the desired object. This is a great way for newcomers to find the more difficult planets such as Uranus and Neptune.

Our geographic location on the Earth determines what we can see and where we need to look. For example, an observer situated along the equator may observe a star directly overhead but to an observer south of the equator this star will appear northward, while the opposite is true for observers north of the equator. These different aspects are due to the spherical nature of our Earth. Since the horizon curves away from us, observers in the Southern Hemisphere are unable to see certain stars, for example, Polaris, near the north celestial pole. The same can be said for observers at northern latitudes looking toward the south celestial pole. Fortunately, the planets are easily visible from most inhabited locations. However, favourable visibility varies to different degrees depending on the time of year and your location north or south of the equator.

## The ecliptic

When using a chart, star wheel or computer program you will find a dotted, perhaps coloured line representing the plane of the ecliptic. The band of constellations through which the ecliptic passes defines the astrological zodiac. The ecliptic is an imaginary line across the sky or celestial sphere along which our Sun appears to travel against the background stars. It is, in fact, the projection of the plane of Earth's orbit around the Sun and is the reference point (0°) by which the angles of orbital tilt in the other planets is gauged. It is called the ecliptic because eclipses of the Sun or Moon can only occur when the Moon passes through this plane.

The orbits of the planets are inclined to the ecliptic plane, however, by very small angles and their subsequent positions in the sky always appear close to the ecliptic. When a planet or comet's orbit crosses the ecliptic plane from south to north it is said to be in *ascending node*, while *descending node* refers to its crossing from north to south. The nodes are the points on the celestial sphere where the plane of its orbit intersects the ecliptic plane.

This imaginary line curving across our sky does not remain in one fixed position. Because the Earth's axis is inclined by approximately 23.5° to this plane, the ecliptic appears to move by 23.5° north and south of the equator during one Earth day or axial rotation. For observers in Australia and other parts of the Southern Hemisphere the ecliptic is highest for night-time observing during the winter months. But the planets are also moving in their own orbits about the Sun and their positions relative to the constellations as seen from Earth change from year to year. Thus, at some point, each planet passes through summer constellations when the ecliptic is unfavourably low to the horizon. The steadiness of an object as seen in the telescope's eyepiece at this altitude is often poorer, since refracting dust particles and air thickness is greater.

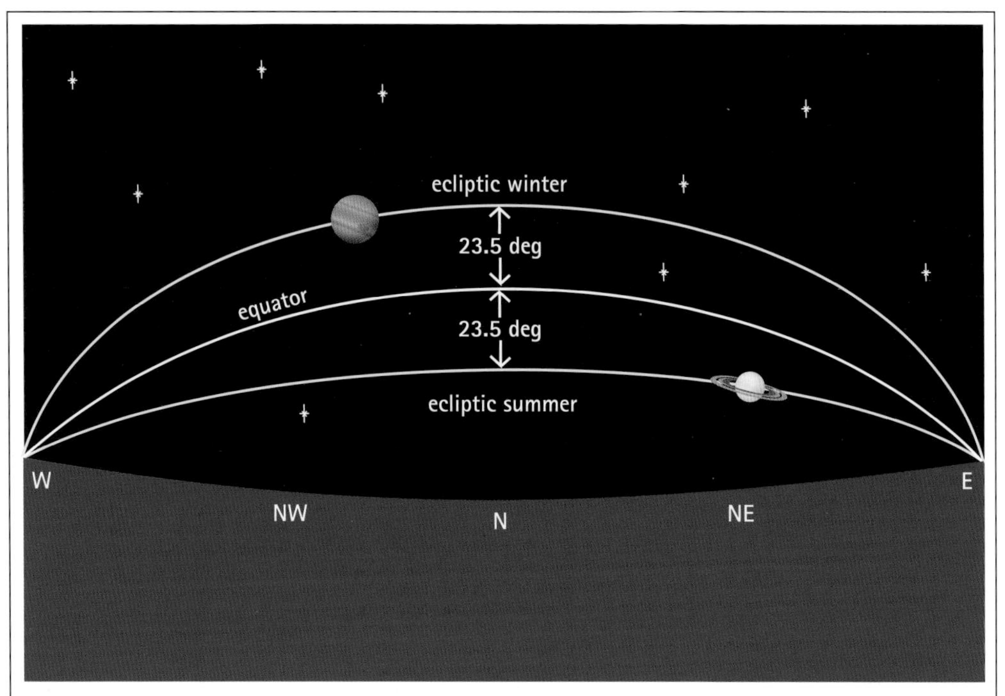


Here we see how the ecliptic path moves north and south of the celestial equator due to the Earth's axial tilt. This view simulates that seen from mid-Southern Hemisphere latitudes around midnight. By midday, when the Earth has rotated through 180°, the situation is reversed.

So the ecliptic serves as a useful benchmark when locating the planets. The outer planets of the Solar System never venture too far north or south of it. Another telltale sign in determining a planet from a star with the naked eye is the steadiness with which the planets often shine compared to the common twinkling appearance of surrounding stars. This is because the planets are so much closer and show a small angular disc while stars are infinite points of light that are more easily distorted by moving air cells high in the atmosphere. But when air movement in the upper atmosphere is highly turbulent, like the edge of a storm front for example, then even the planets will twinkle erratically.

Six of the planets can be observed without the aid of a telescope. The most prominent in terms of brilliance are Venus, Jupiter, Saturn and Mars. Mercury's proximity to the Sun and visibility low on the horizon subdues its potential brightness, also limiting observable time. Distant Uranus is extremely faint and in certain circumstances borders the performance limits of the unaided eye. Of the remaining planets, Neptune requires at minimum the aid of binoculars to locate it and a star chart will greatly assist positive identification. Pluto is very faint and requires a large aperture telescope and detailed star chart to positively identify it among the stars.

## Motions of the planets

Before we venture on to the telescopic characteristics of the planets we should first look at their motions across the sky from one apparition to the next. An *apparition* is the period of time when a planet is visible in the sky and is not consumed by the glare of the Sun. The inner planets Mercury and Venus are called inferior planets and have shorter orbital periods. The outer planets (those outside the orbit of Earth) are called superior planets and their orbital periods take longer with respect to their individual distances from the Sun. The orbital distance and speed of each planet determines how quickly it appears to move through the constellations of the zodiac. A superior planet is said to be at *conjunction* when it is situated on the opposite side of the Sun as seen from Earth. It is said to be at *opposition* when Earth is situated between it and the Sun. An inferior planet is said to be at *superior conjunction* when it is situated on the opposite side of the Sun in relation to Earth and at *inferior conjunction* when it passes between the Earth and the Sun.

Let's now follow the course of a superior planet's apparition. Once the planet has passed through conjunction you can rise in the early hours before sunrise and start scanning the eastern dawn sky with the naked eye for Mars, Jupiter or Saturn. Rising a little earlier each day, the planets become notably brighter as they creep further from the Sun's glare and into the early morning darkness. Once each planet starts to rise around the middle of the night it is at right angles to the Sun and said to be at *western quadrature*. This is the time when a superior planet exhibits its greatest phase effect and a narrow strip of its night side can be seen along its western limb. Mars presents the most obvious effect since it is closest to Earth, however Jupiter also appears less than fully illuminated. Since Saturn is even further out than Jupiter, its disc shows little obvious signs of phase or night side, but as its rings appear so elongated we can observe the shadow of the planet's disc cast across them with dramatic almost 3D-like views. This in itself can keep one mesmerised for hours.

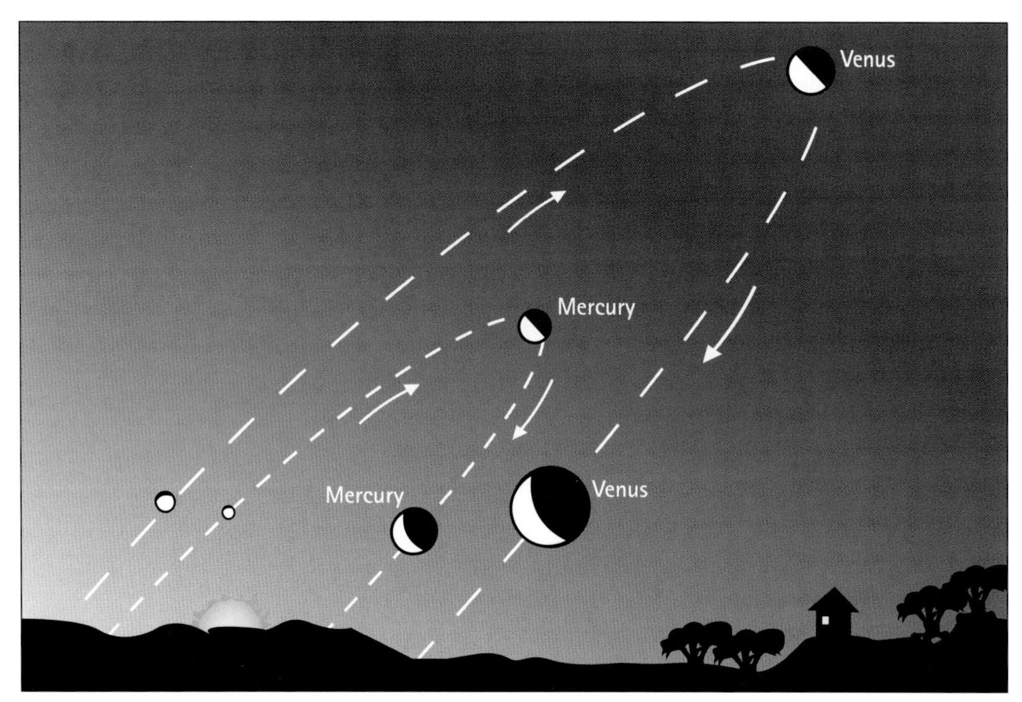


After passing through superior conjunction we can observe the progress of the inner planets in the western evening sky. The apparent angular size of the planet increases as its phase decreases along the narrowing path between it and the Earth. After reaching greatest separation from the Sun it eventually slips back towards sunset.

As our planetary apparition progresses we find our familiar friend now rising around sunset.

Like a car race with Earth catching up on the inside lane, we find the gap is narrowing. As we monitor the planet's slow eastward drift with respect to the background stars it appears to stop. Earth has now caught up with the planet in the outside lane and this is referred to as that planet's *stationary point*. The planet then appears to drift backward, or with retrograde movement, in a westward direction for a short time until it reaches yet another stationary point. This apparent backward motion is an illusory effect as seen from our perspective and during this period the planet reaches opposition, when it is said to be situated on the opposite side of Earth to the Sun. After reaching a second stationary point, the planet then resumes an easterly drift once Earth has passed it on the inside lane. If you were to plot its position with respect to the background stars you would notice that the planet appears to make a small loop in the sky. The closer the planet is to Earth, for example Mars, the more dramatic is the effect. During opposition, a planet's disc appears fully illuminated and since the distances between us are at their minimum, it is also the best time to observe.

Once a planet has passed opposition it then continues to rise earlier in the day until it is situated again at right angles to the Sun in the sky. This is called *eastern quadrature* and a phase effect of the planet's disc is seen once more, however this time on its eastern limb. As the weeks pass by, the time gap between sunset and the setting of our planetary wanderer narrows further until it becomes too difficult to detect in the dusky glow after sunset. It reaches conjunction once more and a new apparition process begins.

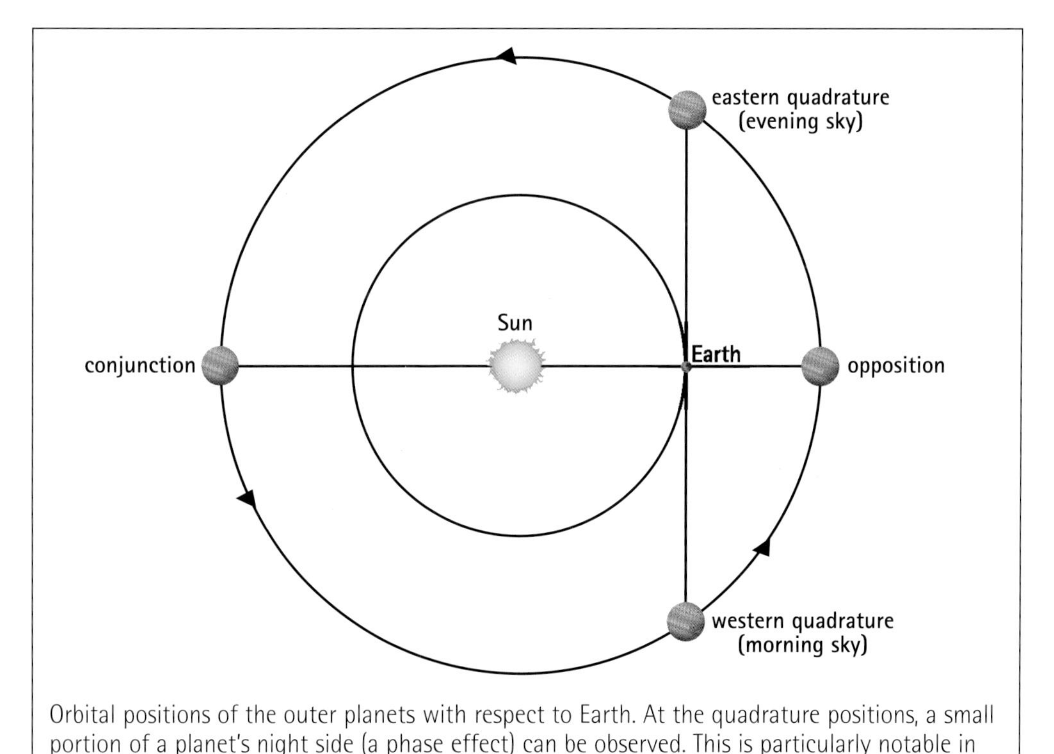


Orbital positions of the outer planets with respect to Earth. At the quadrature positions, a small portion of a planet's night side (a phase effect) can be observed. This is particularly notable in the case of Mars and to a lesser extent in Jupiter.

# Seeing clearly—your observing site

One of the great benefits of lunar and planetary observing is that we don't have to travel great distances for dark skies. Aside from the most distant and the fainter asteroids, our neighbouring celestial partners are bright, distinct objects that can be viewed from most urban areas with little or no difficulty. You can even observe from the balcony of an apartment building, providing you have a reasonably unobstructed view to the ecliptic. For ultimate convenience and comfort, your own backyard or housebound domain is the preferred option from which to observe, yet it may not always be the best. If plagued by tall surrounding buildings or large leafy trees you may need to travel to a better location.

If at all possible, avoid setting up your equipment on hot asphalt or concrete driveways, or aiming at targets over warm rooftops. Both these scenarios can dramatically degrade the image at the eyepiece due to the turbulence caused by heat rising from these surfaces. Such restrictions, however, may not always apply during colder months of the year when daytime heating of these surfaces is milder. Best practice is to observe during the cooler hours of the evening when external surfaces have acclimatised to the surrounding air.

## Permanent observatory

If you have a good clear view of the ecliptic throughout most of the year, you might consider erecting a permanent site to house your telescope and equipment in the backyard. Commercially available fibreglass domes come in a range of sizes and some are fabricated here in Australia. These are miniature versions of those seen on mountaintops but without all the internal staircases! A more economical approach might be to purchase a do-it-yourself garden shed or modify an existing shed to include a roll-off or flip roof. A home observatory is a great way to store your telescope on a permanently polar aligned pier, thereby negating the need to carry out this task each time you observe. It also preserves your carefully collimated optics from potential misalignment due to inadvertent bumps during transportation or set up.

If your home observatory is large enough, you can also house your computer, video equipment and other cameras. All equipment cabling and power requirements can remain conveniently intact, making your observing experience even more enjoyable and productive. Let's face it, if there is any chance that clouds may roll in at the last moment, the television can often win an argument with that little voice in our heads when pondering the thought of dragging all the gear outside and setting it up.

## Obstructions to a clearer view

Unlike the purity of the vacuum of space, we live on the surface of a planet surrounded by a protective blanket of moving air. To some extent we can liken it to being submerged in clear water, where the view overhead sometimes ripples and shimmers. A view of the Moon or a planet can appear to go in and out of focus like a slide projector. When we observe the planets we want to use the maximum practical power our telescope can theoretically produce so as to see as much detail as possible. Air turbulence, sky transparency and magnitude all affect how much detail can be gleaned at the eyepiece.

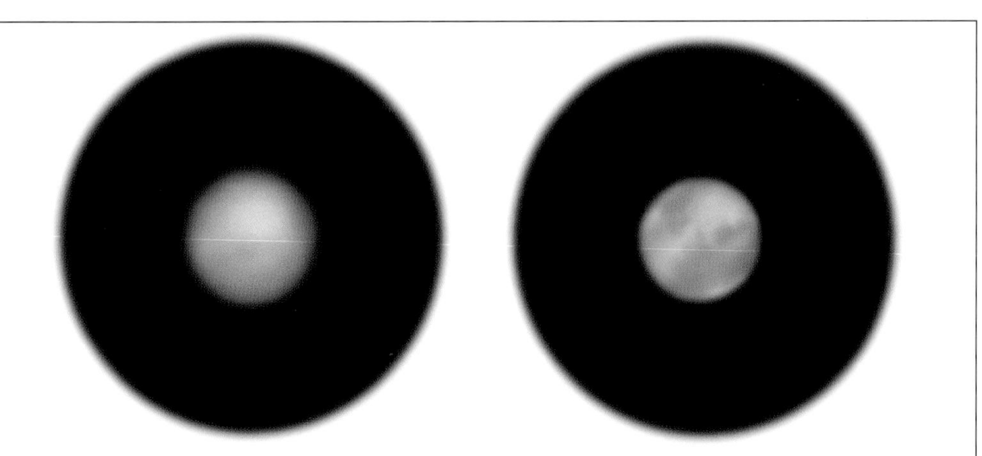

Turbulence in the atmosphere and local convection currents can greatly affect what is seen through a telescope, even if the sky is perfectly clear. Features on the Moon or the planets appear to ripple like running water. In these views from a 200mm (8in) Newtonian telescope, the surface details of Mars are severely blurred at left. At right, 'the seeing' was far better several hours earlier and the observable features showed greater contrast.

### Air turbulence

Most of us are familiar with the childhood song *Twinkle, Twinkle Little Star*. A twinkling star has a certain romantic overtone, somewhat like gazing at a full moon rising over the ocean, or a glorious sunset. However, for astronomers the song might well go, 'Twinkle, twinkle little star, how I wish you'd settle down!'

Stars twinkle because the light they emit is being distorted by turbulence in the air. The amount of distortion affecting the view is referred to as *the seeing* and can be quantified and noted using the simple scale below:

1. perfect seeing—no shimmering even at high powers
2. good seeing—slight undulations with periods of steadiness lasting several seconds
3. moderate seeing—large tremors
4. poor seeing—constant, troublesome undulations
5. very bad seeing—constant look of boiling, with little or no useful detail (i.e. poor contrast), blurry even at low powers.

During the day, both the Earth's atmosphere and surface are heated by the Sun's rays. Through the process of convection, air currents, jet streams and turbulent eddies are produced near the surface and at mid to high altitudes. Their presence is most notable at the eyepiece of a telescope, particularly at high powers. These pockets or streams of dense air create the boiling or rippling effect seen when observing a star, a planet or even the craters on the Moon, subsequently distorting fine details and reducing image contrast. They are ever-present, yet to varying degrees. It is for this reason that observatories are built on high mountaintops where the atmosphere is thinner, and moreover why the United States spent US$2.3 billion putting the Hubble Space Telescope into orbit.

The level of turbulence will soon determine whether you're in for a long night of observing or an early night to bed. Sometimes, conditions can change dramatically throughout the course of the night, and at others only for a brief moment of time. Depending on the severity of 'poor seeing' you can try stepping down to an eyepiece yielding a lower power to reduce the noticeable rippling in the image. The best observing times are considered to be just after sunset, before sunrise or around midnight when the variation of temperature with altitude is minimal. No matter how we look at it, twinkling stars are bad news for planetary observers but a fact of life we must contend with.

## Sky transparency

Though not nearly as critical to planetary observers as the seeing, sky transparency is another factor that can affect the amount of discernible detail. Light scattered through particles suspended in the atmosphere reduces both the apparent blackness of the background sky and observed differences of contrast seen in subtle planetary details. A thin haze overhead will reflect city or suburban light pollution downward while the light of a full moon will be scattered throughout. Sky transparency can mean the difference between visually spotting a faint moon of Saturn or the presence of a small dust storm on Mars. Sky transparency is usually at its best after heavy rainfall or strong winds when city smog, bushfire smoke or dust has been washed out of the sky or blown out to sea. It is usually rated on a scale from 0 to 10, whereby 0 is completely overcast. A rating of 10 is near perfect and stars of the 6th magnitude can be seen with the naked eye. Even if sky transparency is poor this does not mean that observing is useless. Some of the most stable seeing I have ever encountered was when sky pollution was high, and with the aid of filters, contrast in detail can certainly be improved.

## Magnitude

*Magnitude* is a logarithmic scale for measuring the brightness of a celestial object. In the simplest terms, the higher the number, the fainter the object. The ratio of perceived brightness from one magnitude to the next is equal to ⅕ the root of 100; therefore a 1st magnitude star is considered 2.512 times brighter than a 2nd magnitude star. A star of the 6th magnitude is 100 times dimmer than a 1st magnitude star. As strange is it may seem, the scale for objects brighter than 0 magnitude moves into negative numbers. For example, Venus at its brightest is –4.2 while the Sun is –27.

There are other more scientific systems for calculating magnitudes of the stars or point sources such as *absolute magnitude* or *photographic magnitude*. For amateur work, however, *apparent visual magnitude* is the standard used for non-stellar objects like asteroids or planets. Typically, the average person can see stars and planets down to around the 6th magnitude in clear dark skies. Less than perfect skies will have a lower magnitude limit—stars will only be visible down to magnitude 5.5, for example.

If you are keeping a log of your observations then it is useful to note the seeing conditions and sky transparency magnitude limit, particularly when sharing your observations with others for comparative analysis.

# Part two: Observing the Solar System

The following chapters are an introductory guide to aid you in seeing more at the eyepiece than can be gleaned from a cursory glance. Indeed, from this starting point you may decide to take on more serious studies in a particular area of interest to you. This is not uncommon among serious amateurs and there are many dedicated groups, Internet resources and other literature available to assist you. Some amateur astronomers choose to focus their attention on the daily behaviour of the Sun or the dynamic changes in Jupiter's atmosphere. You might decide to take part in a patrol for the detection of dust storms and other phenomena on Mars, or you might simply be content to notch up the number of personally observed planets and photograph them. Whether it's the Sun, Moon or the family of planets, each amazingly individual target will offer many exciting and observational challenges to keep you occupied for years. The hobby is only as limited as the challenges you set yourself.

# The Sun

The creator of life and eventually our ultimate fate, the Sun is the most prominent object in the sky. Pulsating at the heart of the Solar System, it is a blindingly bright and relentless inferno of colossal proportions. The Sun is the most influential component in our existence. In everyday life, its presence on a clear and windless day can give us a positive feeling of comfort while on an overcast day we can feel perhaps a little less inspired. As morning breaks or the day is drawing to a close, a beautiful sunrise or sunset holds a certain romantic appeal for most of us. Without its life-giving energy all living things on the surface of the Earth would perish. Beyond this basic awareness most people don't give it much more thought as they go about their daily rituals, taking its presence for granted.

It is amazing that even today most people are astounded to learn that our Sun is in fact a star—the nearest star to Earth. They are almost as equally astounded when learning that the stars are indeed other suns, but at much greater distances. This simple realisation is often enough to inspire newcomers to the hobby with a sudden and almost overwhelming new sense of the enormity of our galaxy and the universe.

Formed at the heart of our Solar System from a collapsed nebulous gas and dust cloud several billion years ago, the Sun eventually ignited as a continuous thermal nuclear explosion. At its core, this enormous nuclear reactor converts hydrogen into helium at temperatures reaching some 15 million degrees Celsius. Imagining such temperatures is almost as difficult as contemplating the distances to the farthest known galaxies. This incredible energy is converted and radiated as heat and light of all electromagnetic wavelengths, some of which are safe, while others are somewhat more dangerous to life. The protective properties of our atmosphere shelter us and have allowed living cells to evolve on the Earth, but some of the dangerous wavelengths (such as UV light) still reach the surface.

Our protective shell is also sometimes violated in other ways. Our Sun is an awesome entity of unimaginable power that can unleash huge amounts of energy in violent eruptions, spewing atomic particles outwards into the Solar System. Taking a day or so to reach the Earth, these particles can disrupt crucial radio frequencies, knock out sensitive satellite equipment and even bring down power grids. Such occurrences are usually followed by increased, and at the same time quite beautiful, auroral activity seen more commonly from far northern and southern latitudes. Most commonly documented is the activity in the Northern Hemisphere called the Northern Lights.

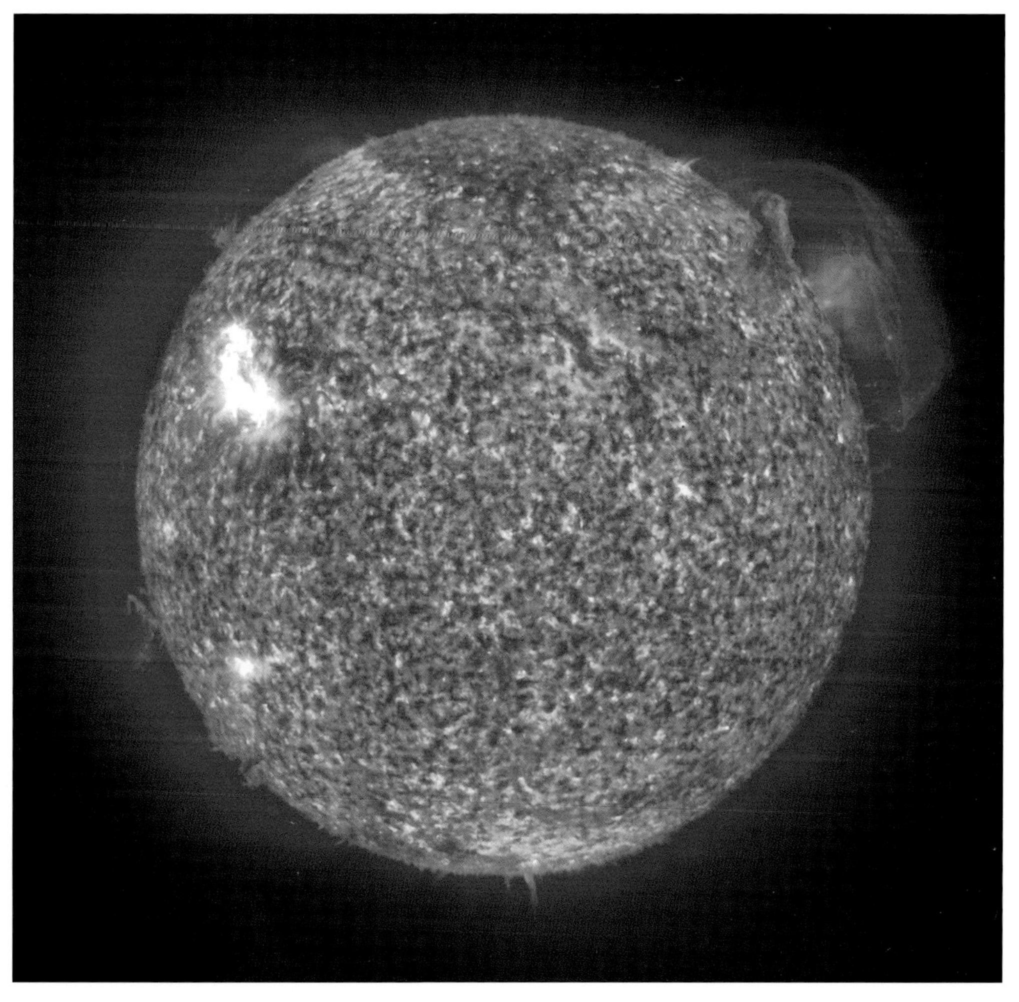

This image shows our Sun, the nearest star to Earth, as seen through the eyes of the *SOHO* spacecraft on 26 August 1997. The hottest areas appear almost white, while the darker red areas indicate cooler temperatures. *(ESA/NASA)*

The Sun is huge, with a diameter of 1 392 000 kilometres across its equator, almost 109 times that of Earth. Like other stars, it is comprised of hot gas and expels about 4 million tons of solar material each second. Huge ejections of gas, in the form of solar flares, can release as much energy as 10 million hydrogen bombs.

Although the Sun has no solid surface, what we see as the visible surface is called the *photosphere*, which has an average temperature of around 5500°C. Surrounding the photosphere is an extremely hot and violent region called the chromosphere. This region can best be seen and photographed during a total solar eclipse, when the fine tentacle-like coronal streamers extend far out into space.

Although the Earth is an inner planet of substantially closer proximity to the Sun compared to the outer planets, the reality of scale and distances beyond our earthly travels becomes evermore apparent when we consider how long it takes the Sun's light to reach us. A room is illuminated almost instantaneously when we flick a switch but the light from the Sun takes more than 8 minutes to reach Earth. If it were to suddenly extinguish, we would not know about it for that amount of time.

# Observing the Sun

As an astronomical target the Sun offers a wealth of interesting and ever-changing features that can be observed with the aid of a suitably filtered telescope. But please always remember: **never observe the Sun directly**. Instant blindness can result!

A sunrise or sunset can be observed with the naked eye because its light has to pass through more of the atmosphere to reach us than when the Sun is high. The brighter blue and green wavelengths are scattered by thicker layers of air and airborne dust particles, giving the more reddish appearance we find so appealing. To watch either event offers an eye-opening display of just how quickly our Earth rotates. Since the Sun's disc spans only 0.5° of sky, the whole sunset event takes around 2 minutes once it first makes contact with the horizon. Like a magic trick done with mirrors, a sunset also demonstrates optical deception. Due to the refractive properties of denser air at the horizon, the solar disc we can see is indeed an optically bent projection of the physical Sun which has in fact already set.

## Observing the Sun with a telescope

To observe the Sun directly, a safe and approved solar filter such as Baader solar film must be used and with great care. These filters are placed over the front, or objective end, of the telescope. When using them, be sure the surface is not scratched or cracked, especially in the case of a glass-coated filter. Never leave the telescope unattended when observing and only allow others to view the Sun under the strictest supervision. If you are concerned about observing the Sun with an aperture filter you might consider the solar projection method instead. This requires the use of a solar projection screen that can be purchased as an accessory or simply made of white cardboard. The light entering the objective is focused onto this screen, which is located at the opposite end of the tube in the case of a refractor. You can even focus the solar disc onto a nearby wall.

Most critical is safe targeting of your telescope. Don't ever look through the viewfinder—in fact, leave the lens caps in place to avoid inadvertently catching a blinding glimpse. Although the finderscope is small, it still greatly magnifies the light and heat from the Sun and can blind you. The safest method is to point your instrument roughly towards the Sun and observe the shadow cast on the ground. Move the telescope about on its mount until the projected shadow of the telescope tube appears circular or as narrow as possible. With a safe solar filter placed over the objective of your telescope, the visible light of the Sun's photosphere and all other wavelengths is reduced by 100 000 times, revealing patchy dark spots called sunspots. Such filters allow safe viewing of the Sun for unlimited periods. Some solar filters produce an image that is white, while others may be orange or yellow. Baader solar film produces a white image, however, using a Wratten No. 23A or 25 filter with the eyepiece can produce a more yellow hue if you prefer.

Viewing the Sun safely requires a solar projection screen or a solar filter placed over the objective of your telescope. Taking advantage of the off-axis aperture stop incorporated in the cap of this Newtonian telescope, a piece of Baader solar film is securely taped to the inside and reduces light transmission by almost 100 000 times. This is a cost-effective solution, however more expensive full-aperture filters are available.

# Sunspots and faculae

Sunspots are cooler regions on the visible disc of the Sun. In reality they are still blindingly bright and extremely hot, but in contrast to the surrounding photosphere they appear dark (through a light reduction filter), almost black by comparison. Sunspots are huge magnetic disturbances within the photosphere and often appear in pairs or groups. Sunspot pairs have opposing magnetic polarities much like a horseshoe magnet. Some are so large they could consume the Earth several times over and can even be seen without the aid of a telescope or binoculars. They can grow over a few days and fade away again within a week or two. On closer inspection you will notice that sunspots are darker at the centre with a lighter or greyish region around the perimeter, sort of like the yolk and the egg white. The dark central region is referred to as the *umbra* while the lighter outer region is called the *penumbra*. Spots have varying shapes but are generally oval or sometimes circular in appearance. Using higher powers sometimes reveals a structure of several smaller and closely situated spots that give the appearance of one large sunspot at low power. As pairs of sunspots rotate into view on the limb of the Sun, the leading spot is called the 'p' or preceding spot while its trailing partner is referred to as the 'f' or follower spot.

Sunspot activity is known to increase and decrease in 11-year cycles. Making a record of sunspot activity is interesting and may be useful to solar astronomers compiling behavioural data. This can be done by simply making drawings of the solar disc and mapping approximate positions and sizes of sunspots over several days, months or perhaps even years. As the Sun has no solid surface, its fluid-like mass rotates at different rates. Around its equator, one rotation takes a little more than 25 days while regions near the poles take up to 34 days. Because the Sun does indeed rotate, you will notice a change of position for each sunspot when compared with your drawing or photograph from the previous day. New sunspots or groups may also appear, so be diligent in your record keeping so as not to confuse one with another.

Another observation of interest that arises from the non-solid nature of the Sun is a phenomenon known as limb darkening near the edges of the disc. More visually subtle and more easily seen near the limbs of the solar disc is the mottled appearance of the surrounding photosphere. This mottling is called *faculae*, and is the bubbling and churning process at the surface of the photosphere. Larger telescopes can resolve this detail into even finer granulation.

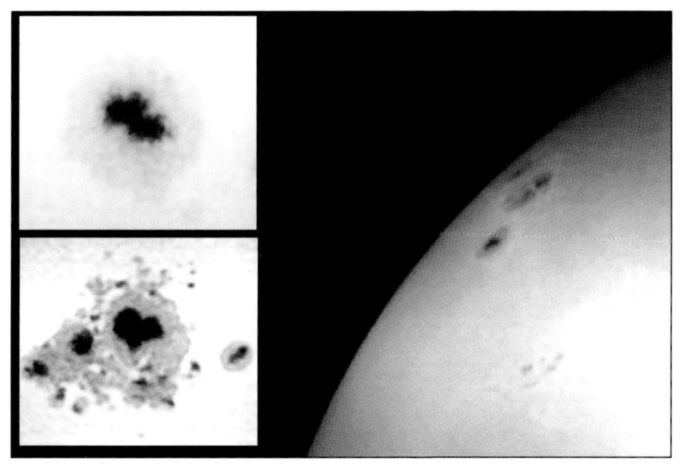

Even a small telescope, fitted with a safety solar filter, reveals faculae and sunspots across the photosphere. The inserts show close-up views of two sunspots captured with video on 19 August 2002. Note the dark central regions of the spots, called the umbra, compared to the lighter surrounding penumbra. The lower insert shows the complexity of a rather large group of sunspots.

## Solar prominences and flares

If you're serious about observing or taking pictures of the Sun then you'll want to purchase a special filter that can block out all wavelengths of light except for the very narrow band emitted by hydrogen atoms. A hydrogen-alpha filter (h-alpha) reveals amazing detail at the 656-nanometre wavelength. Huge prominences can be seen arching tens of thousands of kilometres into space and sometimes looping back onto the solar disc. Flares and filaments on the solar disc are revealed with great detail. Some flares are extremely violent and can reach out to a frightening 100 000 kilometres or more into space.

To witness the Sun through such a filter is awe inspiring, and being mindful of its immense power and size is also very humbling experience. H-alpha filters come in various

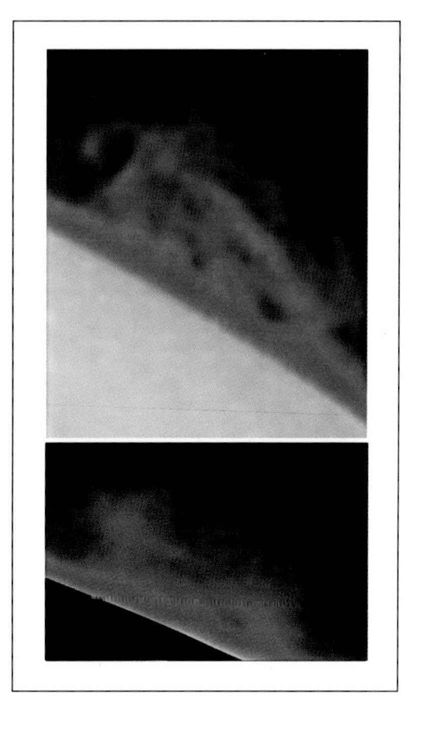

These two images were taken about 20 minutes apart with a camcorder held up to the eyepiece of a Schmidt Cassegrain telescope which was fitted with a hydrogen-alpha filter. In the bottom view the solar disc has been cut away during image processing to produce an occulted disc effect.

sizes and are proportionally more expensive with increased aperture. Filters for refracting telescopes with apertures of 50mm (2in) to 100mm (3.9in) are commonly available, however if your dealer doesn't hold stock one may have to be imported from the United States.

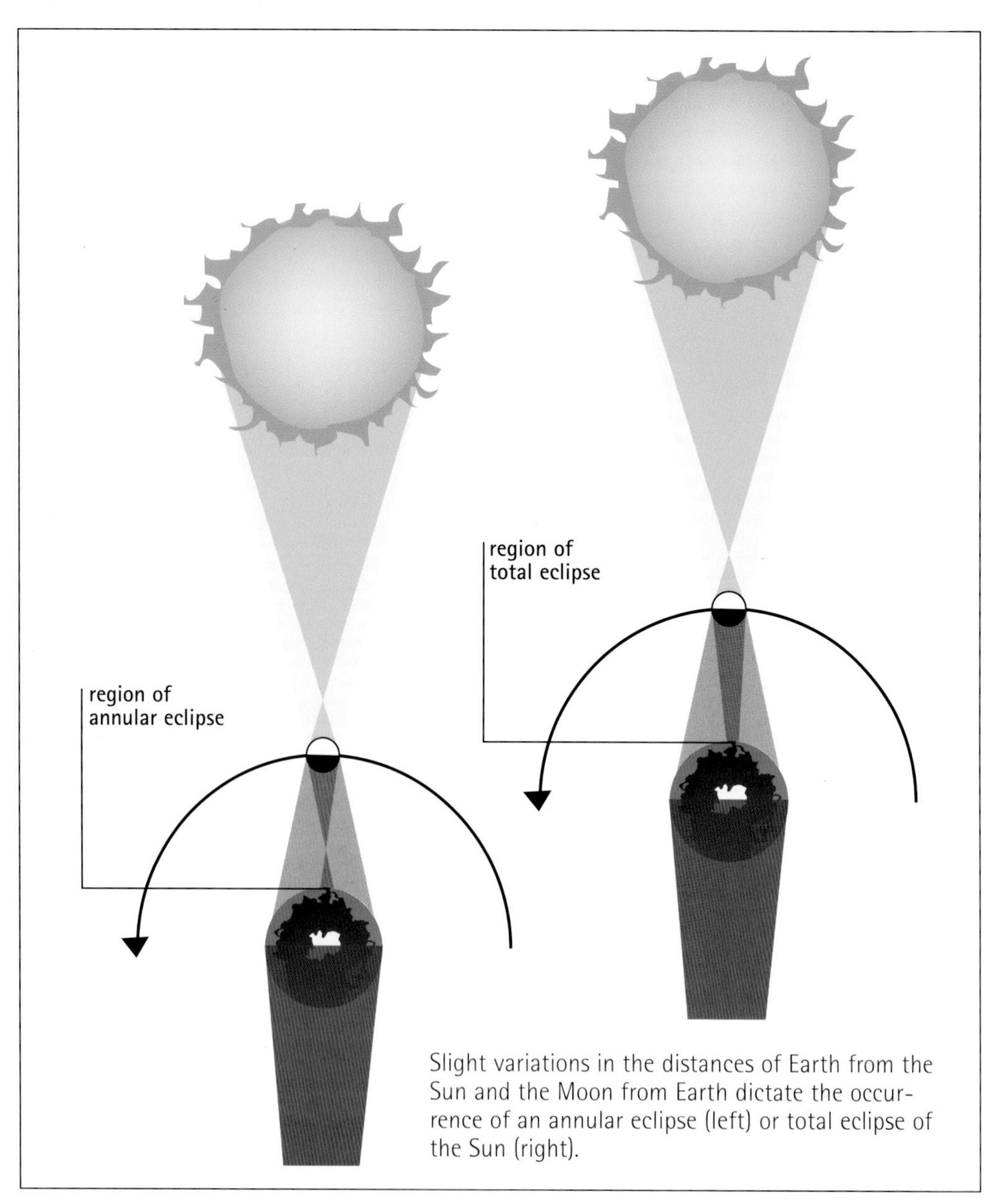


Slight variations in the distances of Earth from the Sun and the Moon from Earth dictate the occurrence of an annular eclipse (left) or total eclipse of the Sun (right).

# Solar eclipses

A solar eclipse occurs when our Moon passes between the Earth and the Sun, crossing over the ecliptic in the sky. A solar eclipse, and in particular a total eclipse, is one of nature's great coincidences in that our Sun is so incredibly large (1 392 000 kilometres in diameter) and our Moon is, by comparison, so insignificantly small (3476 kilometres in diameter). Yet due to the Sun's enormous distance from Earth (around 150 000 000 kilometres) its apparent diameter appears equal to that of our Moon, which is a mere 384 000 kilometres from Earth. So perspective takes over and both take up roughly 0.5° of sky.

## Partial eclipse

Like the name infers, a partial eclipse occurs when the Moon passes across only a portion of the solar disc. During these events it is interesting to study the dips and peaks of crater rims, mountains and valleys along the lunar limb when it is so highly contrasted in silhouette against the bright solar disc.

A partial eclipse of the afternoon Sun on 16 February 1999. Such events can be captured using a simple camcorder and the solar film from a pair of eclipse shades placed across the front of the camera lens. The Sun grows progressively redder as it sinks towards the western horizon.

## Annular eclipse

The orbits of the Moon around the Earth and the Earth around the Sun are not circular but slightly elliptical. Therefore the angular diameters of the Sun and Moon as seen from Earth vary slightly as their distances from the Earth change. When the Moon's angular diameter is less than that of the Sun during an eclipse, the outer ring of the Sun's photosphere is seen and this is called an *annular eclipse*.

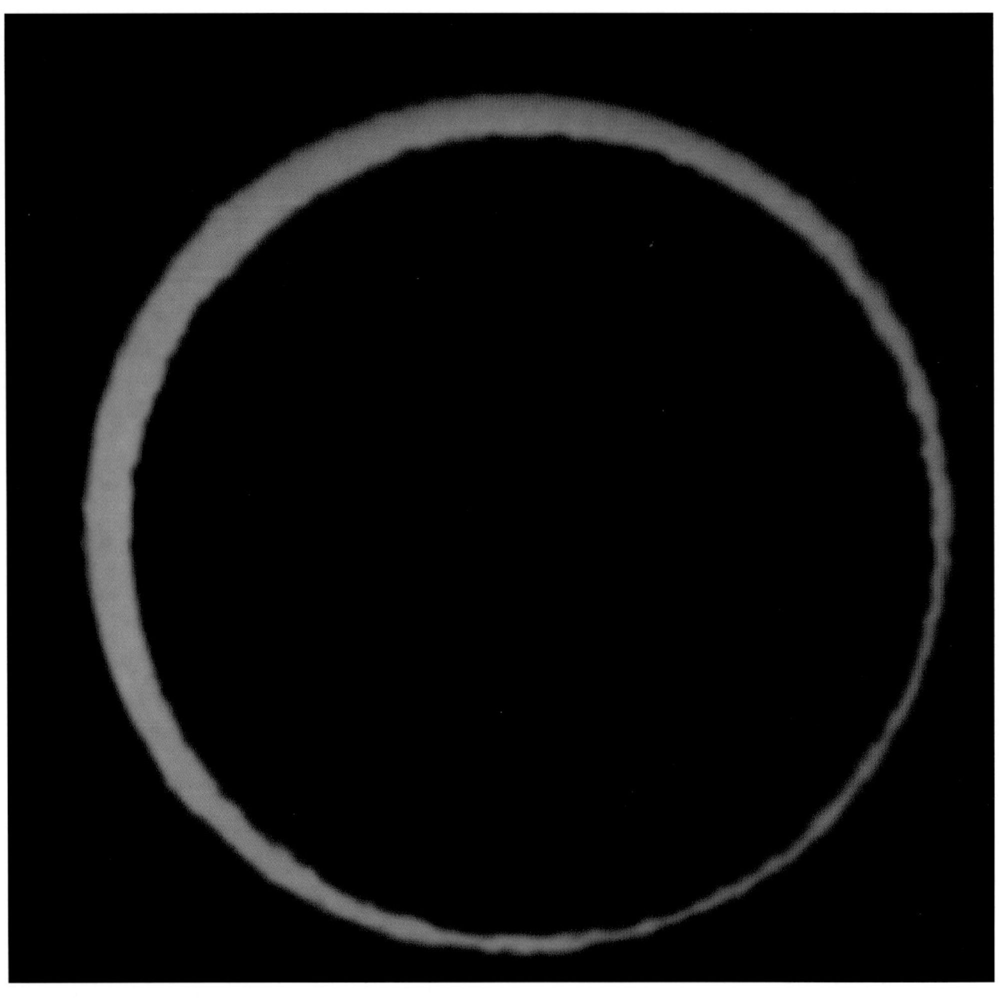

During a solar eclipse, when the Moon's angular diameter is slightly less than that of the Sun due to varying orbital distances, you can see the outer ring of the Sun's photosphere in an event called an annular eclipse.

## Total solar eclipse

The most popular and exciting event of all is a total solar eclipse. This presents a chance for professional and amateur astronomers alike to witness and study the delicate outer corona. Relative to the much cooler outer regions of the solar disc, the temperature of the corona (its outer atmosphere) can reach as much as 2 million degrees Celsius. Eclipse chasers, as they are sometimes called, will travel around the globe in order to capture each event from the best possible vantage point.

This beautiful multiple exposure was taken near Lake Everard in South Australia during the total eclipse of 4 December 2002. Using a medium-format camera and ISO 50 film, exposures were taken at 5-minute intervals, from $^1/_{250}$ of a second at f/22 to $^1/_{30}$ of a second at f/8 for the partial phases, and $^1/_4$ of a second at f/4 for totality. *(Courtesy Logan Shield)*

The most recent event in Australia took place on 4 December 2002 in Ceduna, South Australia. Starting in the Atlantic Ocean just west of Africa, the skies began to darken as the Moon commenced its passage across the face of the Sun. Its shadow, or umbra, passed along a thin strip that proceeded in a north-easterly direction through the Great Australian Bight and across South Australia, covering an area just over 260 kilometres wide. Totality at Ceduna lasted little more than 30 seconds, however some total eclipses can last several minutes.

As totality comes to an end, a beautiful diamond ring effect unfolds as the Sun's photosphere emerges from behind the Moon.

The next event in Australia is not due to occur until 14 November 2012 but you don't have to wait this long to see another total eclipse. If planning an overseas holiday at some time in the future, why not try and work it around an upcoming eclipse event and make the trip even more memorable? Popular astronomy periodicals advertise organised eclipse tours throughout the world so you'll have plenty of time to plan.

## Taking pictures of an eclipse

A total solar eclipse is most definitely a photo opportunity not to miss. Taking video, digital stills or photographic film pictures of the Sun requires the same type of solar aperture filter that you would use with your telescope. The filter can be removed during totality when the glare from the Sun's disc is completely blocked by the Moon. To create a picture so that the Sun takes up a good portion of the frame will require a telephoto lens up to 1000mm. Digital and video camera users should only utilise the optical zooming capability of their cameras. Digital zoom features do not produce any additional detail beyond the camera's optical limit. Since most common digital still and video cameras offer little more optical zooming capability than 18X, an optional telephoto adaptor is a worthwhile investment.

An alternative option to directly imaging the Sun while the Moon makes its passage across the solar disc, leading up to totality and after totality, is to take pictures of the event as projected onto a solar projection screen then turn your camera to the Sun during totality.

Your camera will achieve best results if mounted on a tripod or a driven mount. Whether using video or a conventional camera, the shutter speed will determine what appears in the resulting picture. The recommended 35mm film speed is ISO 100, which will produce less grainy pictures than faster films. Prominences extending from the blackened disc during totality will show up better with shorter exposures around 1/500 of a second at f/8. Reducing the shutter speed, thereby increasing exposure time, will reveal more of the bright outer corona. The recommended exposure time here is about 1/30 of a second. If using a video camera, make sure you set the auto-focus function to 'off' and manually set it to infinity in order to avoid annoying in- and out-of-focus images. Use the manual shutter speed (usually an incrementing push button or thumb wheel) to adjust for optimal exposures.

The beauty of video is the ability to see what is happening in the viewfinder or the LCD monitor as it happens. You can make adjustments accordingly and quickly. If you are inexperienced and using a film camera, it is best to bracket your exposures (i.e take numerous photos with different exposure times) in quick succession in order to improve your chances of obtaining at least one shot at the perfect exposure.

The total solar eclipse of 4 December 2002 is just underway in this camcorder image taken from Sydney. Using a 3X telephoto adapter and the camera's full optical zoom, a piece of Baader solar film was used to prevent overexposure and damage to the camera's CCD chip.

# The Moon

| | Moon | Earth |
|---|---|---|
| Mean angular diameter (arc min/sec) | 31m05.2s | - |
| Equatorial diameter (km) | 3 476 | 12 756 |
| Surface gravity (metres per second$^2$) | 0.162 | 9.78 |
| Magnitude (brightness at full moon) | –12.7 | - |
| Max distance from Earth (km)—apogee | 406 700 | - |
| Mean distance from Earth (km) | 384 401 | - |
| Minimum distance from Earth (km)—perigee | 356 400 | - |
| Number of known moons | - | 1 |
| Orbital inclination (degrees) | 5.2° | 0 |
| Synodic period (days new moon to new moon) | 29.53 | - |
| Primary atmospheric composition | - | nitrogen/oxygen |
| Rotational period (d/h/m) = sidereal month | 27.32 | 0h23h56m |

While the Sun dominates the daytime sky, the Moon is the prominent celestial feature of the night sky. Of course our views of the Moon are not just limited to dark hours. The Moon can be seen and observed through a telescope during the day, but contrast is somewhat subdued. The closest celestial body to Earth, the Moon has captured our imagination since the dawn of humankind and it is not surprising that it should become our first heavenly conquest. Postwar rocketry advanced quickly when the two superpowers of the time, the United States and Russia, raced each other in a bold attempt to land a space probe on the surface of this ghostly sphere in the sky. The first successful missions were simply photographic fly-bys and our first view of the lunar far side was radioed to Earth from the Russian probe *Luna 3* in October 1959. These pictures revealed a heavily cratered and mountainous terrain very different from the near side. In fact, the far side of the Moon almost completely lacks

the large solidified lava plains so prominent in the hemisphere that faces Earth. Two dark regions of the far side were named Mare Desiderii (Sea of Dreams) and Mare Moscovrae (Sea of Moscow).

Among several probes to follow, the United States sent their probe *Ranger 9* plummeting into the lunar surface while beaming back photographs of its own descent into the crater Alphonsus. As *Ranger 9* drew ever closer to its demise, the pictures revealed a constant unfolding of craters within craters, from tens of kilometres deep to just centimetres wide. Russian probe *Luna 9* would later make the first successful soft landing in February 1966, but while the Russians concentrated most of their efforts on robotic probes, the United States were well under way with their own plans of landing a person on the Moon. And so it happened on 20 July 1969 when Neil Armstrong made his epic first few steps on lunar soil. After *Apollo 11*, and with the lunar program ending in the 1970s, the five missions that followed brought a plethora of new knowledge to scientists here on Earth, increasing our then relatively limited understanding of the Moon and its make up.

Whether orbiting the Moon in a spacecraft or observing it through a telescope from the backyard, the most obvious features are its enormous craters. For some time it was held that the origin of these circular impressions was perhaps volcanic. However it was later mooted that they might be the result of huge meteoric impacts and this is now known to be the case. Nearly all of the craters we can see are impact scars left over from the final bombardment of the Solar System as each planet and moon swept up the remaining rocky debris from its creation. Since the Moon has no protective atmosphere it is constantly peppered by micrometeorites even today.

The lunar surface or 'soil' is called *regolith* and is comprised of layers of rock and other materials excavated from large meteoric impacts. Smaller impacts over millions of years continued to refine the granularity of this regolithic material, like

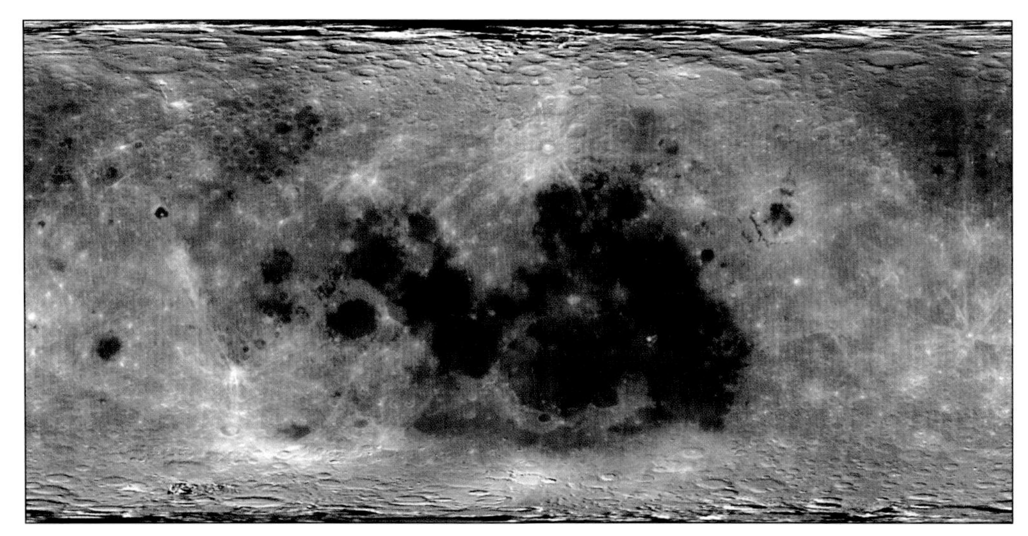

A global map taken by the *Clementine* spacecraft, which mapped the Moon with four cameras in unprecedented resolution in 1994. Note the abundance of dark maria confined to the centremost region, being the side of the Moon that always faces Earth. *(NASA/Naval Research Laboratory)*

turning the soil in a garden but with a sandblasting effect. The lunar dust is very dark and is likened to that of finely powdered graphite.

Nearly half a ton of rock samples returned to Earth have revealed that the Moon is essentially made from the same material as the Earth. This led to the theory that it may have formed from a collision between Earth and a Mars-sized world, leaving a ring of debris orbiting Earth that eventually coalesced to become our Moon. Since the Moon is much smaller than Earth, with an equatorial diameter of only 3476 kilometres, it is also less massive and has a much lower *escape velocity*. Escape velocity has to do with the relative speed at which an object, such as a rocket, must travel at in order to leave the gravitational influence of a larger body, such as the Earth. Earth's escape velocity is 11.2 kilometres per second while the Moon's is only 2.4 kilometres per second. This makes our neighbour an excellent low-fuel-consumption launch pad for future space exploration. While a full moon appears extremely bright here on Earth, it is actually a poor reflector of light due to the properties of its dark regolith and basalt surface. The light from a full moon measured here on Earth is around 0.25 *lux* (lux being the unit of measurement for illuminance) while a full Earth as seen from the Moon is about 16 lux and it takes 1.3 seconds for reflected sunlight from the Moon to reach Earth. Surface temperatures on the Moon vary greatly from between –170 to –184°C at night to +130°C at maximum during the lunar day. We have it pretty good here on Earth!

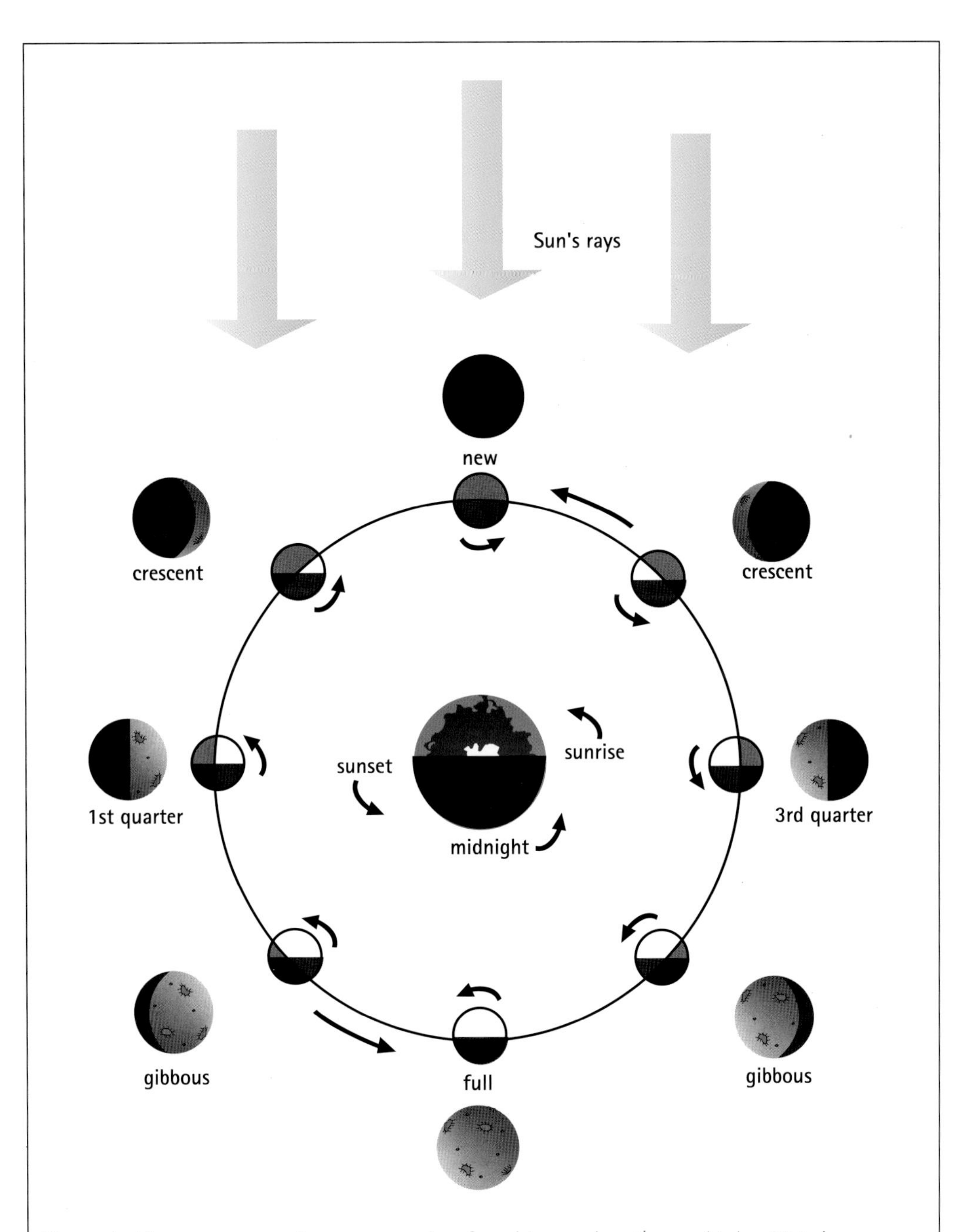


The period from new moon to new moon is referred to as a lunation and takes 29.5 days. This diagram represents the view of an observer hovering high above the Earth's north pole. The outer views of the Moon represent how its phases would appear as seen from Earth.

# Observing the Moon

Our nearest celestial neighbour offers earthbound observers the most detailed views of another world and is always a favourite when it comes to seeing striking detail through a telescope. Along with some influence from the Sun, the Moon is the main controller of the ocean tides of Earth. Aside from well documented effects on the behaviour of various sea and nocturnal creatures, some believe the Moon's influence to be more far-reaching, even affecting people's moods around times of a full moon. Like a golden sunset, a full moon is considered an object of romantic appeal, but it has also been tied to scenes from classic horror movies. Orbiting our planet at an average distance of 384 401 kilometres, the Moon takes 27.3 days to complete one revolution with respect to the background stars—this is referred to as its *sidereal orbital period*. However, from new moon to new moon (a *lunation*) the period is 29.5 days. During each orbit, the Moon's distances from Earth change slightly since it does not revolve about us in a perfect circle. This slightly elliptical orbit means that at its closest approach (356 400 kilometres) the Moon appears slightly larger in the sky and this is referred to as *perigee*. At its farthest distance (406 700 kilometres) the Moon is said to be at *apogee* and presents a smaller disc in the sky.

An easy target for any beginner, the Moon is a joy to observe through even the most humble telescope. A good pair of binoculars will easily reveal its vast cratered terrain, mountainous highlands and low-lying dark plains. In fact, you can see smaller features on the Moon through a pair of binoculars than you can see on any of the planets through some of the largest telescopes. For example, the smallest features on Mars which are visible from Earth are about 20 kilometres across while on Jupiter they're as big as Australia. On Saturn they're the size of Africa! But even a 150mm-aperture telescope will reveal craters on the Moon a little over 1 kilometre in diameter and linear features like rilles several hundred metres wide!

Since the Moon rotates on its axis in the same time it takes to orbit the Earth once, we only ever see one half of its globe. In other words the same side faces us all the time, a phenomenon known as *synchronous rotation*. Synchronous rotation is common to moons orbiting other planets of the Solar System. As the Moon waxes from a thin crescent to first quarter and full moon, then wanes to last quarter and beyond, observers have the opportunity to view its surface features under a wide variety of lighting conditions. This ever-changing show of light and shadow provides new and interesting views for even the most seasoned observer.

There is, however, another variable that provides additional perspective to our views of the Moon's earthward hemisphere; this phenomenon is known as

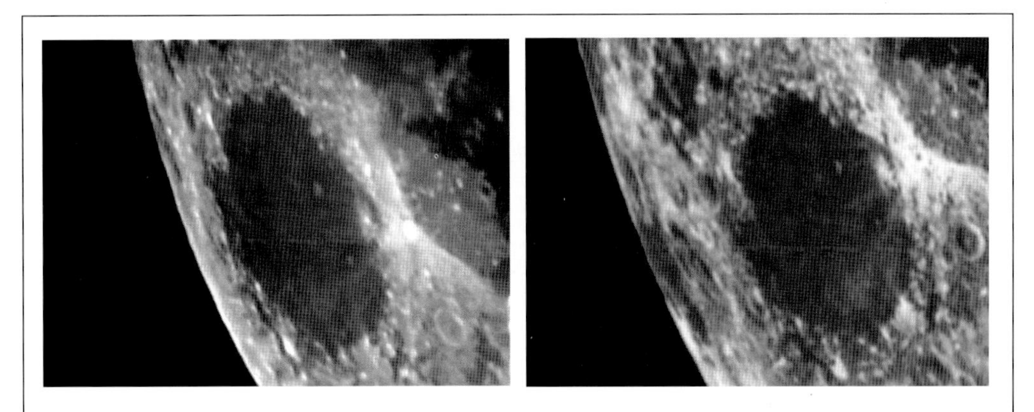

These images clearly show the foreshortening of Mare Crisium near the Moon's eastern limb due to the effect of librations in longitude.

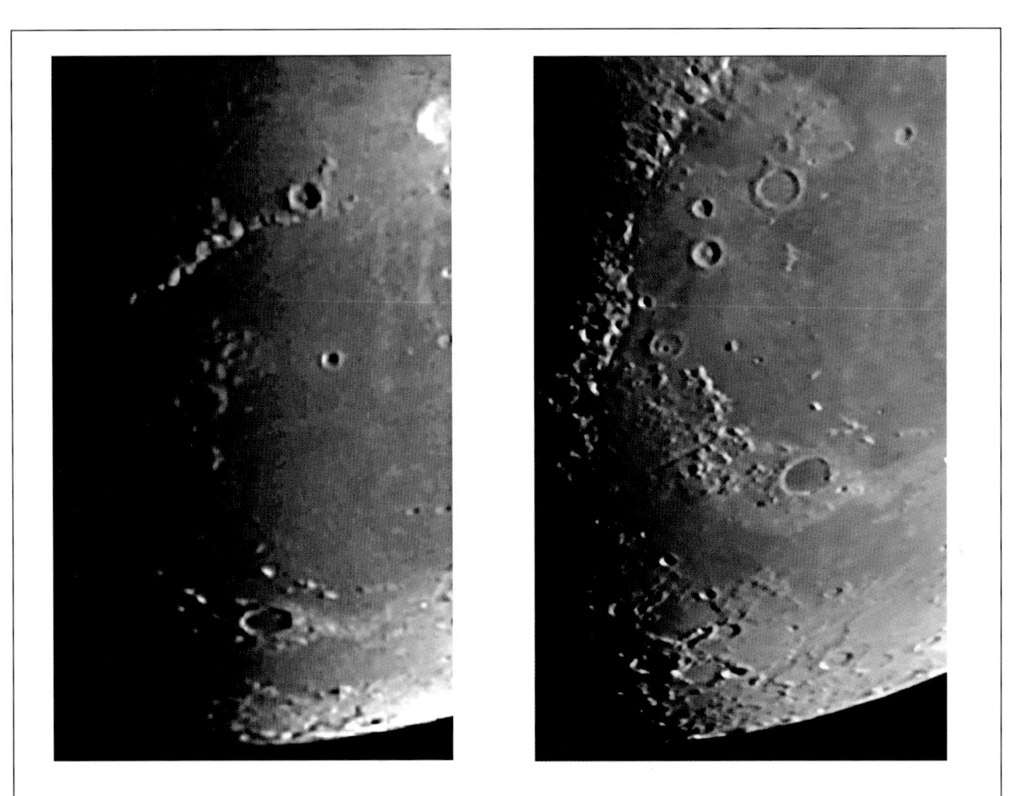

Libration in latitude can also be seen; at left, Plato is somewhat lower with reference to the lunar limb and more foreshortened than it appears in the image at right.

*lunar libration*, the apparent 'wobble' exhibited by the Moon as it circles the Earth. Lunar libration offers earthbound observers the opportunity to examine up to 59 per cent of the lunar surface rather than only 50 per cent.

There are several types of libration that affect how much additional lunar surface we can see, however the greatest effect can be seen from librations in longitude and latitude. The phenomenon of libration in longitude occurs due to the changing velocity of the Moon in its elliptical orbit, and when combined with its constant rate of rotation on its axis we are afforded a peek of a few extra degrees at its eastern and western limbs. As the Moon approaches perigee, its orbital velocity increases. This means that the Moon now passes through 90° of its orbit around the Earth in slightly less time than it rotates 90° on its axis. Approaching apogee, the Moon's orbital velocity slows and its axial rotation appears to catch up and advance in the opposite direction. A corresponding 'libration in latitude' is caused by the 6.4° inclination of the Moon's axis of rotation to the plane of its orbit. This means the Moon reveals extended views of its northern and southern polar regions when it is at opposite ends of its orbit.

## Earthshine

An interesting phenomenon most readily seen a few days before and after a new moon is the ashen light effect on the otherwise dark side of the Moon. Complementing the slender crescent phase, the entire lunar disc can be seen. It is best observed in the darker hours before sunrise or after sunset. This ashen light is caused by reflected sunlight from the Earth, known as *earthshine*. Just as a full moon lights our surroundings during the night, so too does a full Earth on the night side of the Moon. As the sunlit side of the Moon increases through each waxing phase, the less obvious the effect becomes until it is no longer detectable. The inverse is the case for a waning Moon. The first and perhaps most obvious reason for the fading earthshine is due to the increased glare from the Moon's sunlit side as it grows each night. The second reason has to do with the complementary phases of the Earth and Moon. If we were to stand on the night side of a crescent moon facing earthward, we would see an incredibly bright, near fully illuminated planet Earth (*gibbous phase*) suspended in the sky. The reflected sunlight from the Earth and its white clouds would subtly illuminate the surrounding lunar surface. As the sun rises higher and the lunar phase grows, Earth's phase is shrinking and reflects less and less sunlight.

Left: Change in apparent size of the Moon from apogee (farthest point from Earth) and perigee (closest) presents about a 10 per cent difference in observed size.

Below: The Moon's phase as seen from Earth complements the Earth's phase as seen from the Moon at the same time. When the Moon presents a crescent phase as seen from Earth, an observer on the lunar surface would see the Earth in gibbous phase. When the Earth's phase is full or near full, it reflects a large amount of sunlight onto the Moon's surface which can be seen as earthshine lighting the night side of the Moon when only a few days old.

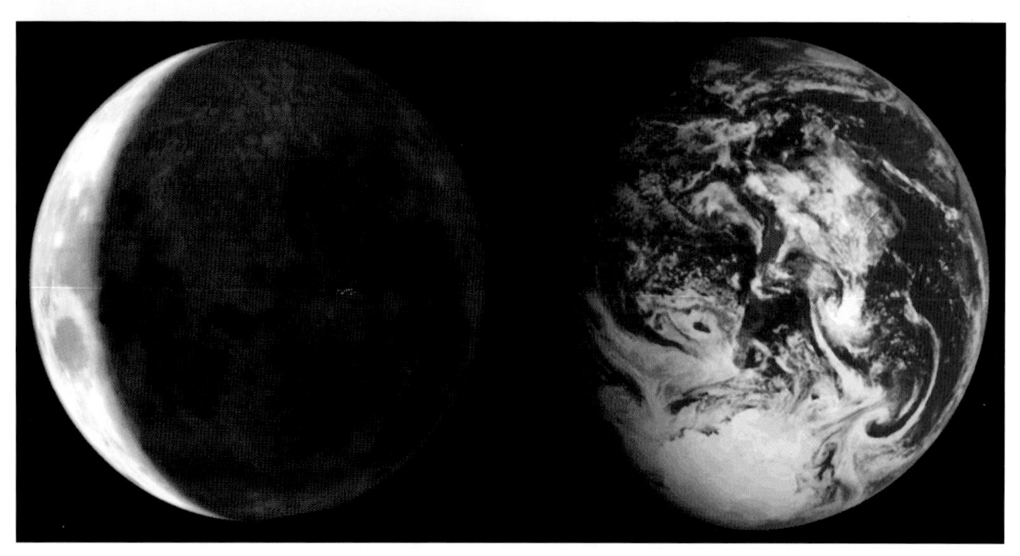

## Interesting targets

The lunar surface is comprised of various landscape features, the most obvious of which can be seen with the unaided eye. When we look up at the Moon we see bright and dark regions. In early times the dark regions were assumed to be great expanses of water, likened to those here on Earth. They were given the Latin names for sea, ocean, bay and lake—*mare, oceanus, sinus* and *lacus* respectively. Later

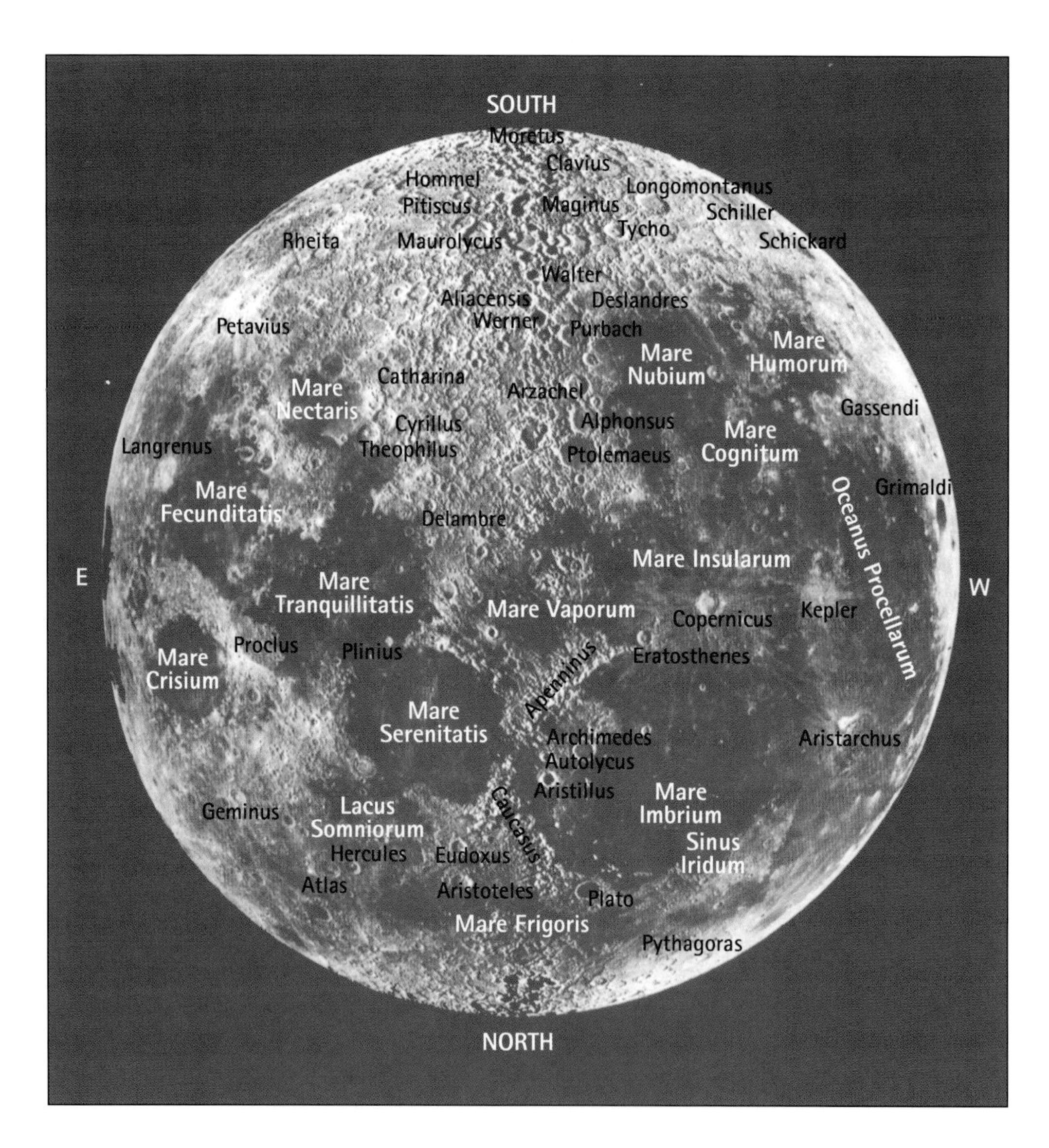


telescopic observation concluded that these regions are in fact barren, dry lava plains resulting from massive impacts during the final violent bombardment of the Moon, when liquid lava seeped up through cracks in the crater floor. The brighter regions are known as the lunar highlands and are much older than the mare basins. Littered with thousands of craters, mountains, valleys and scarps, the lunar highlands make for the most fascinating and diverse telescopic observations you can imagine. Even with a simple 60mm telescope there is plenty to see.

The most effective way to observe lunar features is by following the progress of the terminator each night, the line that borders the dark and lit sides of the Moon. The phase or age of the Moon (the time that has elapsed since the last new moon) determines where the terminator will be. Here we'll look at the most prominent features seen as the terminator sweeps across the Moon. Although most of the descriptions are based on telescopic views, many of the larger craters can be seen with 10X binoculars. To identify and examine the myriad of additional surface features not defined here you will need a good lunar atlas.

Some other terms you will come across when reading about the Moon are:

Catena—a chain of craters
Dorsum—a wrinkled ridge
Mons—mountain (can apply to lunar domes)
Montes—a mountain range or grouping of mountains
Rille—a fissure or channel in the lunar surface, also referred to as a rima
Rupes—scarp or crustal fault
Vallis—valley.

## New moon to first quarter

Observing the Moon a day after new moon is always a rewarding challenge. However, since it is situated so low in the west after sunset and the atmosphere is thicker, telescopic views can appear to boil and bubble, distorting delicate detail. One can attempt to photograph the Moon just after sunset when it presents the thinnest crescent. Since observations of the Moon are most commonly undertaken a few days into the new lunar month the aspects it presents during days one and two can be quite interesting. With a diameter of 570 kilometres, the craggy, mountainous eastern wall of Mare Crisium seems quite alien under this lighting and its lava-filled floor lies partially submerged in darkness, revealing subtle undulating ridges. Immediately north of Mare Crisium, the rim of the 126-kilometer crater Cleomedes reflects the light of the lunar morning sunrise.

2-day-old crescent Moon.

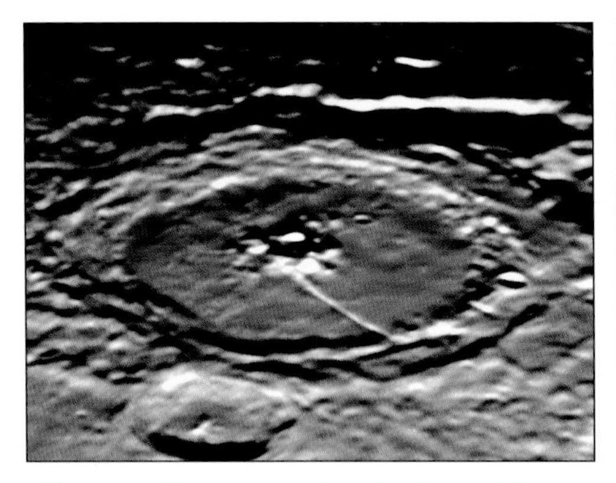

The magnificent crater Petavius is 177 kilometres across, with a range of central mountain peaks and long rille along its floor.

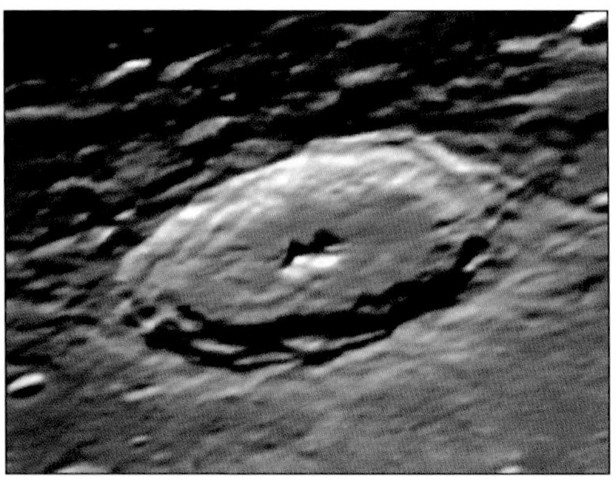

Langrenus, 132 kilometres in diameter, reveals two prominent shadow peaks on its crater floor.

By day 3, roughly 30° and 40° south of Mare Crisium respectively, magnificent views of the large craters Langrenus and Petavius are revealed. Petavius has a magnificent cleft that runs from its multi-peaked central mountains to the crater's rim.

Days 4 and 5 reveal a large, heavily degraded crater from the earliest impact period called Janssen. A complex array of smaller impact craters can be observed within and around its rim. On closer examination you will notice that the entire northern wall has been destroyed by other, later impacts. The largest rille within Janssen is roughly 4 to 6 kilometres across and can be seen through most small

Janssen, a massive and heavily bombarded crater, is almost 190 kilometres across.

Rheita Vallis cuts through the southern lunar highlands for a distance of almost 448 kilometres. The crater Rheita lies partly submerged in shadow immediately to its left.

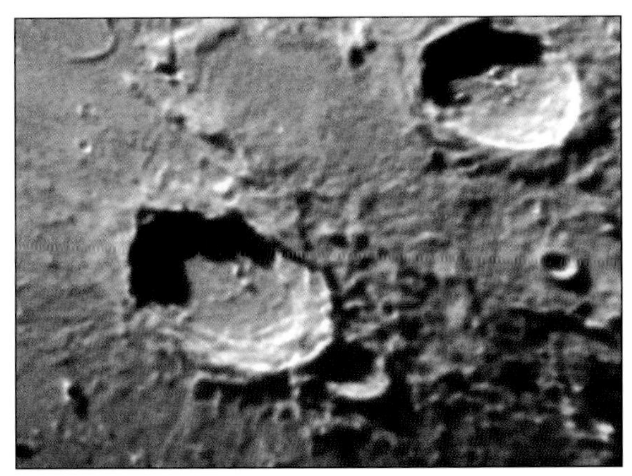

Between a 6- and 7-day-old moon, a striking pair of terrace-walled craters to observe are Aristoteles (lower) and Eudoxus.

telescopes. To the east of Janssen is the very prominent Vallis Rheita. Mare Nectaris, sometimes called the Nectaris Basin, is a large, dark plain almost 900 kilometres across. Under low lighting angles it reveals several rings and ridges. As the lunar sunrise progresses, the massive lava plain Mare Tranquillitatus reveals its eastern section, where a small 12-kilometre crater, Cauchy, can be seen between two parallel crustal fault features—Rima Cauchy, 210 kilometres in length, and Rupes Cauchy, roughly 120 kilometres in length. Rupes Cauchy casts a stunning shadow across Mare Tranqullitatus and becomes a bright feature during lunar sunset. To the north, in a moderately dark plain called Lacus Somniorum, are two rather striking craters side by side. Atlas is roughly 87 kilometres across with a system of fine clefts on its crater floor while smaller Hercules, 68 kilometres across, has a distinct circular crater on its darkish crater floor and another smaller one embedded in its southern rim.

On day 5, 82-kilometre Pitiscus can be found at 54°S and 31°E among a very busy region of highland craters including Rosenberger, Vlacq and the much larger and highly disturbed (fractured and pitted) crater, Hommel. Moving further north along the lunar terminator, a spectacular overlapping pair of craters steals the show. Theophilus is the more prominent of the two, with very distinct crater rim walls rising to around 4350 metres and a broken array of central mountain peaks reaching almost 1400 metres in height. Neighbouring Cyrillus

Apenninus Mountains bordering the Mare Imbrium basin.

A dramatic view of craters Werner (centre) and Aliacensis (upper left) that form an outstanding pair about 8° south-east of Arzachel.

appears much more degraded. Both craters are around 100 kilometres in diameter. Further to the north, on the northeast rim of Mare Serenitatis, a crater called Posidonius presents an interesting array of rimae across its crater floor, with two small craterlets known as Posidonius A and B.

On day 6 another striking pair of craters in the north come into view. Aristoteles and Eudoxus exhibit classic terraced walls inside their respective rims and only small central mountain peaks. In the much busier southern highlands at around 42°S and 14°E, Maurolycus is a very prominent 114-kilometre crater with several smaller impact scars on its floor. The meteor that excavated it replaced an older crater and part of its wall can still be seen to the south.

6- to 7-day-old Moon.

By day 7, or first quarter, the Moon has travelled little more than one quarter of its orbit around the Earth. To the north, Mare Serenitatis is now in full view. The massive eastern rim of the giant Mare Imbrium basin reveals itself as a complex range of

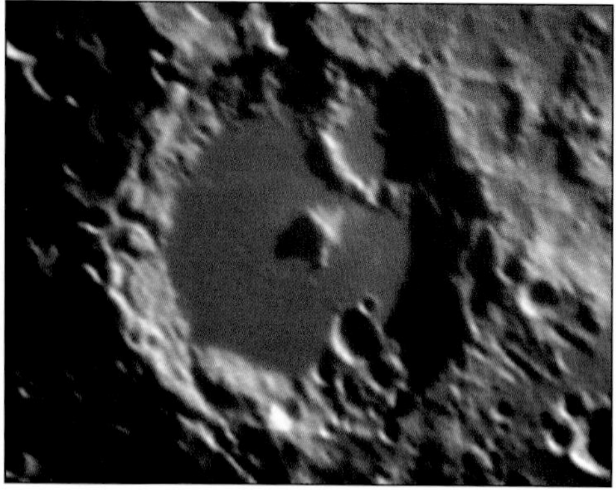

Albategnius, a 136-kilometre crater with a central peak.

mountains. The two ranges of Montes Apenninus and Montes Caucasus are stunning to observe, casting enormous shadows across the vast Imbrium plains. Farther north are the Montes Alps (sometimes simply referred to as the Alps) where a deep, dark gorge known as the Alpine Valley cuts through the mountains into Mare Frigoris. Moving southward to around 6–10°N, along a flat plain bordering Sinus Medii and Mare Vaporum, is a long, shallow valley partly formed by a chain of craters called Rima Hyginus. A small 10-kilometre crater, Hyginus, lies roughly midway along this valley. Both features can be seen quite clearly at 40X power with a 100mm (3.9in) refractor.

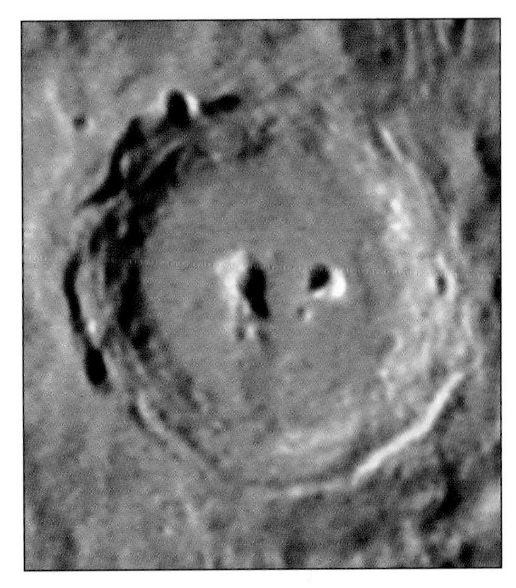

Arzachel, a prominent, well-formed 97-kilometre crater with a central peak.

At 11.2°S and 4.1°E, using a lunar map as a guide, you'll find a magnificent crater called Albategnius which has a diameter of 136 kilometres from rim to rim. A little farther south a nice pair of craters can also be seen; Werner and Aliacensis are 70 and 80 kilometres in diameter respectively. Werner exhibits distinct terraced walls with a rim-to-crater floor depth of roughly 4200 metres. Farther south again, two craters, Stöfler and Licetus, appear most eye-catching while the surrounding highland terrain presents an enormous array of finer detail even in most small telescopes.

From days 8 to 10 a host of new and interesting features unfold. Between days 7 and 8, about half way along the terminator, an array of three large, almost adjoining craters take centre stage among the elite of the most recognisable lunar features. Southernmost is 97-kilometre Arzachel. This crater has well-formed terraced walls and smaller craters on its floor. It also has an interesting central peak, which formed from rebound bedrock under the crater floor caused from severe compression by the impact. Immediately north is Alphonsus, where the probe *Ranger 9* crash-landed in 1965. It also has a central peak and a system of creek-like channels aptly named Rimae Alphonsus that runs along its floor. Immediately north of Alphonsus we see slightly larger Ptolemaeus which is just over 150 kilometres across with a flat crater floor and degraded walls. All three make for an interesting photograph during this lunar phase.

Collosal Clavius, 225 kilometres across, has numerous craterlets on its floor.

In the southern highlands, the massive crater Clavius can be seen a little farther south of prominent ray crater Tycho. Clavius spans more than 225 kilometres and over such a distance its curving crater floor becomes quite apparent at low Sun angles. Several smaller craters are entrenched on its floor and under various angles of illumination many smaller craterlets can be detected but good seeing is required.

To the north are several interesting features given greater prominence due to their locations in the large yet relatively barren Mare Imbrium basin. Archimedes is perhaps the most prominent, with a diameter of 83 kilometres. It has a flooded crater floor where lava seeped up through cracks in the mantle after it was excavated by the impacting meteor. It has an unusual tapering to its terrace-walled rim. Nearby is a conspicuous crater pair, Aristillus and Autolycus, while a little further north is the crater Cassini. In this region two solitary mountain peaks will easily catch your eye. At 41°N and 1°W, Mons Piton rises 2250 metres above the surrounding plains with a 25-kilometre diameter at its base. At 46°N and 9°W, Mons Pico rises nearly as high as Piton. Both cast long witches-hat shadows across the Imbrium basin and if you observe the Moon throughout the night you can witness these shadows shrinking as lunar daylight encroaches.

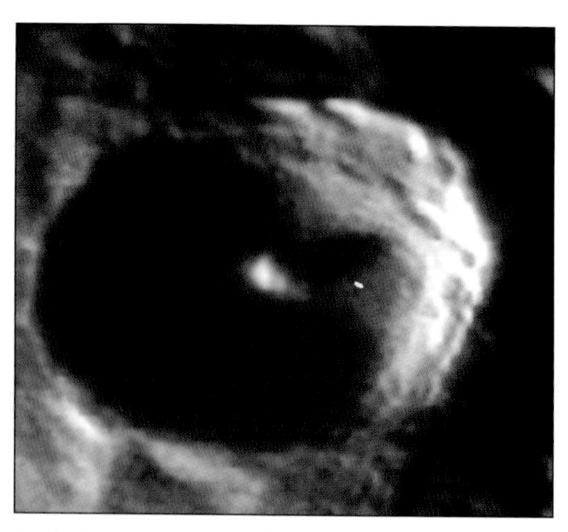

Tycho is a comparatively fresh-impact 85-kilometre crater with a vast ray system spanning almost 1500 kilometres across the lunar surface. It can be seen with the naked eye around times of full moon. The ray system is composed of ejecta material excavated from the impact site.

In the south-eastern section of Mare Nubium, near a small but distinct crater called Birt, you will see a linear feature called Rupes Recta, commonly referred to as the Straight Wall. Roughly 280 metres high and extending for about 110 kilometres in length, this scarp was created by tectonic activity in the lunar crust late in its formation. Poking out like a lollypop on a stick at the end of the Apenninus Mountains bordering Mare Imbrium and Mare Insularum is a well-formed terrace-walled crater, Eratosthenes.

Plato can be seen nestled in the craggy area west of the Alps between Mare Frigoris and the Imbrium basin. Under steady atmospheric conditions and low Sun angles one might spot the tiny 1- and 2-kilometre sized craterlets in its dark flat floor. A telescope with good resolving capability will be required to meet this challenge. This 100-kilometre crater is also an interesting target when it comes to shadow projections across its basin floor. Low Sun angles cast long shadow spires from the tops of Plato's undulating crater rim that can be seen retracting as the Sun rises over several hours.

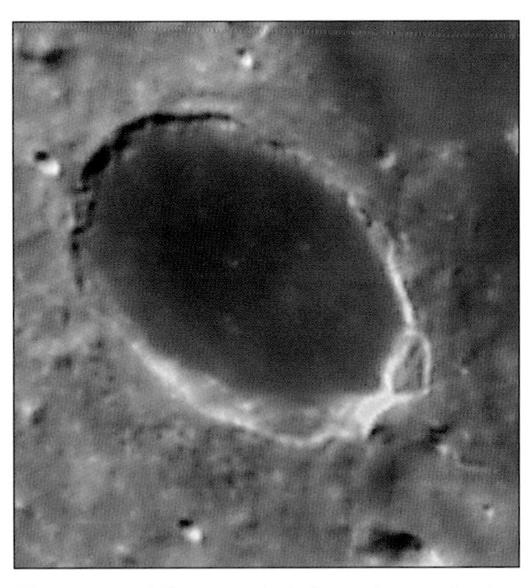

Plato, a 100-kilometre dark floored crater in the highlands south of Mare Imbrium. Peaks along its crater rim cast striking long shadow spires across its floor at low Sun angles. Small craterlets 1 to 2 kilometres in diameter can be seen on its dry basalt floor although with some level of difficulty in smaller telescopes.

By days 9 and 10 we see the magnificent crater Copernicus, bordering Mare Imbrium and Mare Insularum. With striking terraced walls and a prominent central peak, Copernicus also has an extensive ray system. Though not as

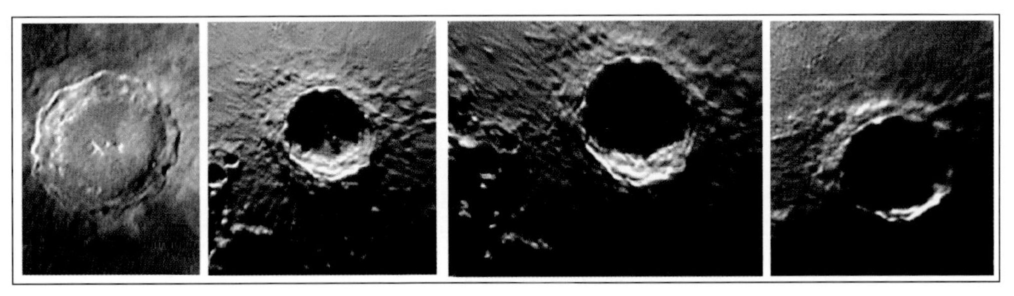

Copernicus is a striking terrace-walled crater. Here it is seen at different stages of illumination or angles of lighting.

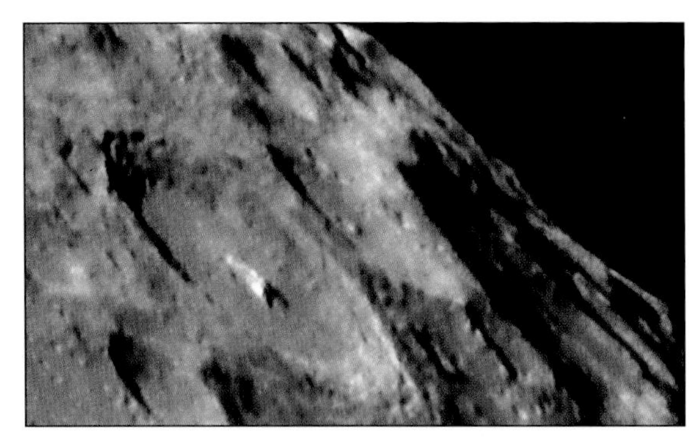

Moretus is a 114-kilometre crater with distinct terraced walls and central mountain peak in the craggy southern highlands with foreshortened appearance.

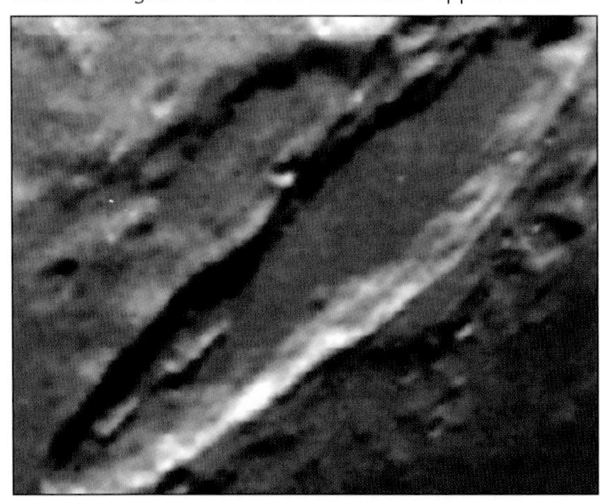

Schiller. Looking more like a crustal fault, this unusual crater is highly elongated with a long ridge along its floor.

Gassendi, a 100-kilometre crater on the border of Mare Humorum.

majestic as the rays of Tycho, they do become more apparent with increased illumination as full moon approaches. At latitude 70°S lies another prominent terrace-walled crater called Moretus. This large crater presents numerous aspects of foreshortening at differing times of libration in latitude. It has a very interesting central mountain peak that casts a long witches-hat shadow during early lunar sunrise.

Between days 10 and 11 near the terminator to the south you will see an unusually elongated crater called Schiller, which measures 179 x 71 kilometres. A long ridge can be seen along its floor as the lunar sunrise progresses.

Moving northward to the Mare Humorum region, a very striking crater encroaches the basin limb. Gassendi is a walled plain spanning some 110 kilometres and features several clefts, central mountains, hills and channels. Further north again a small ray crater called Kepler can be found in the eastern part of Oceanus Procellarum. At 36.6°N and 40.5°W we find an interesting uplifted feature called Mons Gruithuisen Gamma in the dark mare plains. More commonly referred to as a lunar dome with a wide base covering around

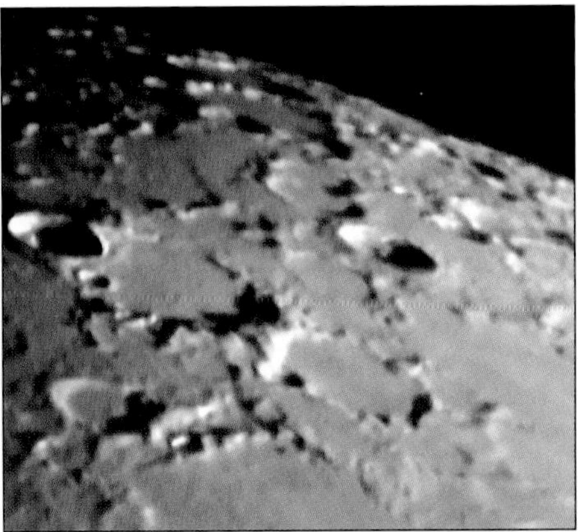

Above left: Lunar dome Mons Gruithuisen Gamma and a nearby crater.
Above right: A favourable libration of latitude and the lunar north polar region is tilted earthward, revealing a highly foreshortened view of the crater Peary just below the polar horizon at centre. Peary is situated 88.6N 33.0E. Craters in the foreground include Anaxagoras, Goldschmidt, Barrow and Scoresby.

20 kilometres, it is best seen under low angles of illumination. With low relief like that of wrinkle ridges (commonly seen within mare basins) the origin of lunar domes is believed to be volcanic.

At days 11 to 12 perhaps the brightest formation under full moon lighting comes into view. Located 23.7°N and 47.4°W, Aristarchus can even be observed on the night side of the Moon, due to earthshine during the first few days after a new moon. Similar in size, and immediately west of Aristarchus, is a plain, flat-floor crater named Herodotus. Extending from this can be seen the longest winding valley on the Moon, called Schröter's Valley. A small 6-kilometre crater is embedded along its narrow path and can just be seen in a small 60mm refractor.

Gibbous waxing moon.

By days 13 and 14 we're skimming the western limb of the Moon for what remains with shaded relief. Near the terminator in the south-western lunar highlands a massive crater called Bailly extends more than 300 kilometres from limb to limb. It is highly foreshortened but can present excellent views during favourable south/west librations. Observing Bailly with high powers and under good seeing conditions can give a sense of standing at the top of one rim and staring across its deep crater floor to the terraced wall of the opposing rim which rises into a black sky. Panning slowly along the lunar limb we see another large 225-kilometre crater called Schickard, situated roughly 20° north of Bailly. Panning even further north to almost 0° latitude we encounter Grimaldi, a foreshortened flooded basin some 172 kilometres in diameter. Just to its west is the crater Riccioli.

When the Moon is 13 to 14 days old, contrast of surface features is greatly reduced. Some craters can seem to almost disappear but the bright ray systems from Tycho and Copernicus take centre stage, as can be seen in this image of a near full moon.

## Full moon to last quarter

By days 14 and 15 the Moon has travelled a little over half way around its orbit of Earth and rises in the east as the Sun sets in the west. The complete earthward face of the Moon is now bathed in sunlight and the craters, valleys and mountain ranges have little or no shaded relief. However, at this time Tycho comes into its own as its apparent, highly reflective ejecta rays are seen spanning thousands of kilometres across the lunar surface and can be seen clearly with the unaided eye.

Like a pair of sunglasses, a neutral density filter or moon filter fitted to your eyepiece is excellent for reducing glare and subject brightness. Colours are unaltered as light is transmitted uniformly over the entire spectrum. Another handy tool is a polarising filter set. Comprised of two filters that thread into either the eyepiece or a filter adaptor, you can vary the degree of transmission. By causing the amount of transmitted light to vary, the filters become, in effect, a continuously variable neutral density filter, which at maximum density limit light transmission to around 5 per cent. These are useful for isolating and examining bright regions such as the crater Aristarchus.

After full moon the lunar terminator reappears on the eastern lunar limb. If you've been following the Moon's progress since day 1 or 2 (lunar sunrise) through first quarter and full moon, you'll now be treated to visually stimulating views of those previously mentioned features under the light of a lunar setting Sun. This is a fascinating new perspective that reveals completely new aspects and detail. The two large southern craters Petavius and Langrenus mentioned previously present perhaps their most striking views when the Moon is 17 days old.

If possible, make the effort to observe the Moon through its waning phases, at least until last quarter when it has travelled three quarters of its journey around the Earth. Since most casual observers follow the waxing phases of the Moon during normal evening hours, many features become familiar over time. However, braving the early morning hours (some might call you a lunatic!) will delight your senses with new perspectives as you witness strikingly different, even alien, views that will astound

Last quarter moon, 21 days old.

18-day-old gibbous moon. After full moon it begins to wane as the night side creeps into the eastern limb. Observing familiar features throughout this part of the lunation cycle provides new and very interesting aspects of lighting and contrast.

you. By day 23 you can officially call yourself a keen observer, or perhaps an insomniac, rising in the early hours of the morning to train your telescope on a slendering Moon that still continues to reveal exciting views under the light of a setting Sun.

Now that you are familiar with the features visible throughout a lunation, you might like to try your hand at sketching or photographing them next time around.

## Lunar occultations

Defined simply, an *occultation* is the eclipsing of one celestial body by another. This might be the Moon passing in front of a star or planet or indeed a planetary occultation with a star or another planet, as was the case with Venus and Mars in October 1590. Technically speaking, an eclipse of the Sun is indeed an occultation of the Sun by the Moon and, to be somewhat more pedantic, an occultation of a planet or star by the Moon is an eclipse.

There is no doubt, however, that our Moon is the most celebrated centrepiece when it comes to an occultation event for three main reasons. From our earthly perspective, the Moon's relatively fast passage of orbit from one lunation to the next and its comparatively large apparent diameter of 0.5° means it sweeps over a greater number of otherwise more distant celestial objects each month. An additional reason stems from the fact that it orbits near the plane of the ecliptic, thus passing over the rich star fields of the Milky Way and occasionally the planets. Our airless Moon is a veritable knife-edge slicing through space, abruptly cutting off the light of distant stars.

Along its west-to-east orbital path across our skies, three main classifications of occultation occur. As it waxes from new moon to near full phases, distant celestial objects disappear behind the night side limb, later reappearing from behind the bright limb. As full moon wanes, an object will slip behind the bright limb, reappearing from the dark limb. These are both referred to as *total occultations*. The third type is referred to as a *grazing occultation*, where a star will appear to skim across the lunar limb, winking on and off as mountainous peaks and valleys pass by.

A scientifically worthy pursuit, information derived from precisely timing an occultation is useful for refining our knowledge of the Moon's orbital elements as well as measuring minute changes in the Earth's period of rotation. If a very close and faint companion star accompanies a bright star, a lunar occultation may block the glare of the bright star, affording a momentary glimpse of the otherwise invisible companion star. To become involved with an organised occultation team you should contact your local astronomy club.

## Occultations of the planets

A close-up telescopic view of the Moon or a planet is one thing, but to capture them together in a dramatic event such as an occultation is quite another and almost certainly something you won't quickly forget. To watch the rings of Saturn peaking over lunar mountains is a breathtaking experience. To gaze at the moons of another planet set on the horizon of our own Moon fills one with an indescribable sense of awe. These are events that can be experienced by anyone with even a small backyard telescope.

Unlike a solar eclipse, when the Moon is directly between the Sun and Earth, a lunar eclipse occurs when the Moon is situated directly opposite the Sun from Earth. At this time the Earth casts a shadow onto the full moon, reducing its brightness substantially. A lunar eclipse can be seen from anywhere on Earth where the Moon is above the local horizon and can last for up to 3.5 hours from first contact to last contact. Due to refraction of the Sun's light through Earth's atmosphere, a small amount of light still reaches the Moon and hence it

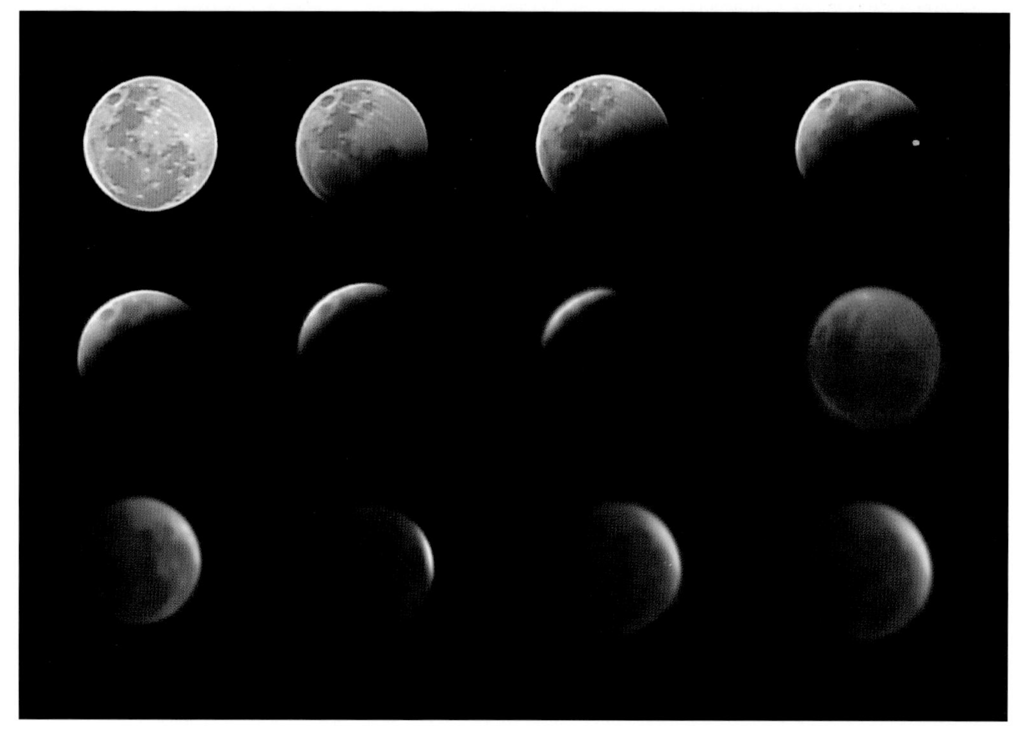

This series of camcorder images shows the progress of the total lunar eclipse on the 16–17 July 2000. The Moon never appears completely darkened during totality but rather exhibits a dark reddish hue, due to refracted sunlight through the Earth's atmosphere.

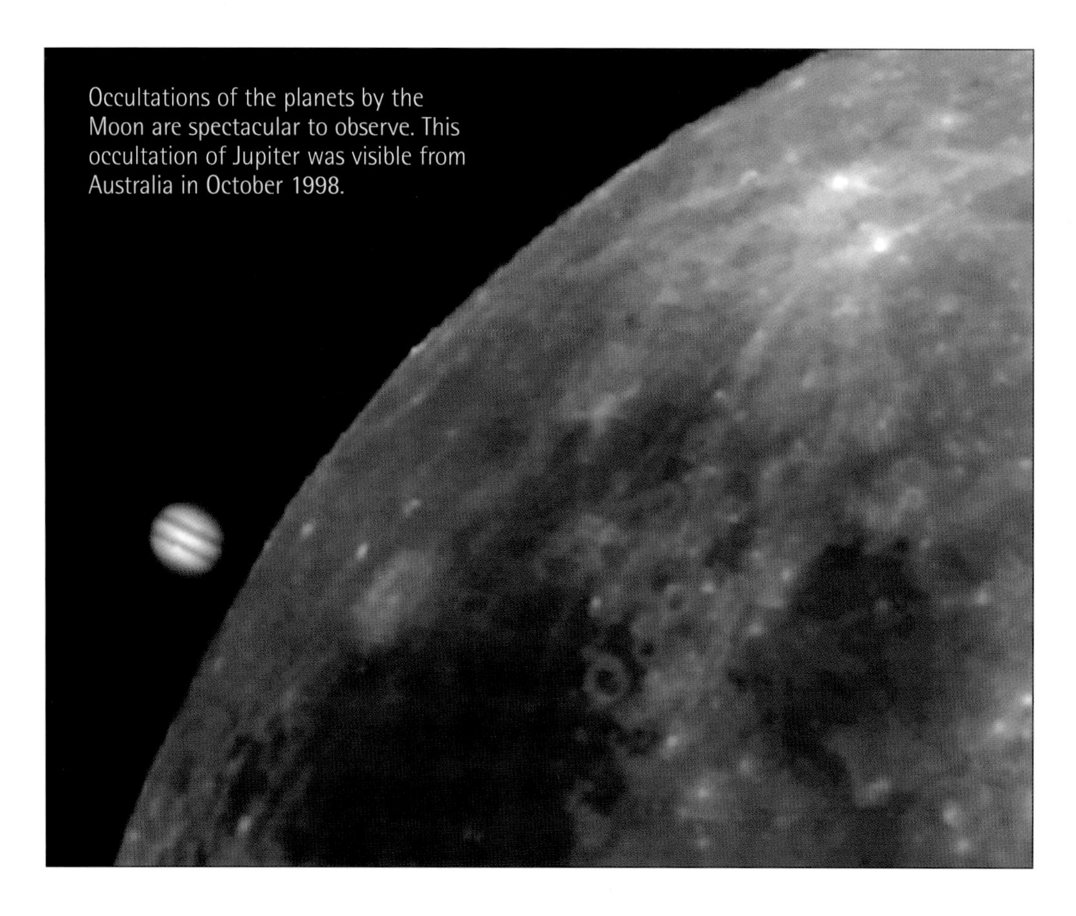
Occultations of the planets by the Moon are spectacular to observe. This occultation of Jupiter was visible from Australia in October 1998.

is never completely darkened during a total eclipse. The Moon takes on a deep red hue during a total eclipse and this is caused by the aforementioned scattering of light through Earth's atmosphere where the component blue light has been filtered out.

As can be seen in the diagram on page 84, the Moon can be immersed in two types of shadow projected by Earth, the outer part being the penumbra and the darker central region the umbra. During a total lunar eclipse the Moon passes completely through Earth's umbra. The eclipse is deemed partial when it is not entirely covered by the umbra, and should it miss the umbra altogether, passing only through the penumbra, it is called a penumbral eclipse.

While it is interesting to observe the curvature of the Earth projected onto the lunar surface during a total eclipse, perhaps the most notable development during totality is the gradual appearance of close proximity faint stars not normally visible under a full moon. If the Moon happens to be passing through a rich star region, such as the Milky Way, during an eclipse event, using your telescope or even a pair of binoculars should enable you to detect several stars being occulted within a few hours.

*Apollo 17* was the last manned mission to the Moon in December 1972. After landing safely in the Taurus-Littrow region, the astronauts stayed for 75 hours as they roamed the surface in a lunar buggy and collected the last hand-picked rock samples for return to Earth.

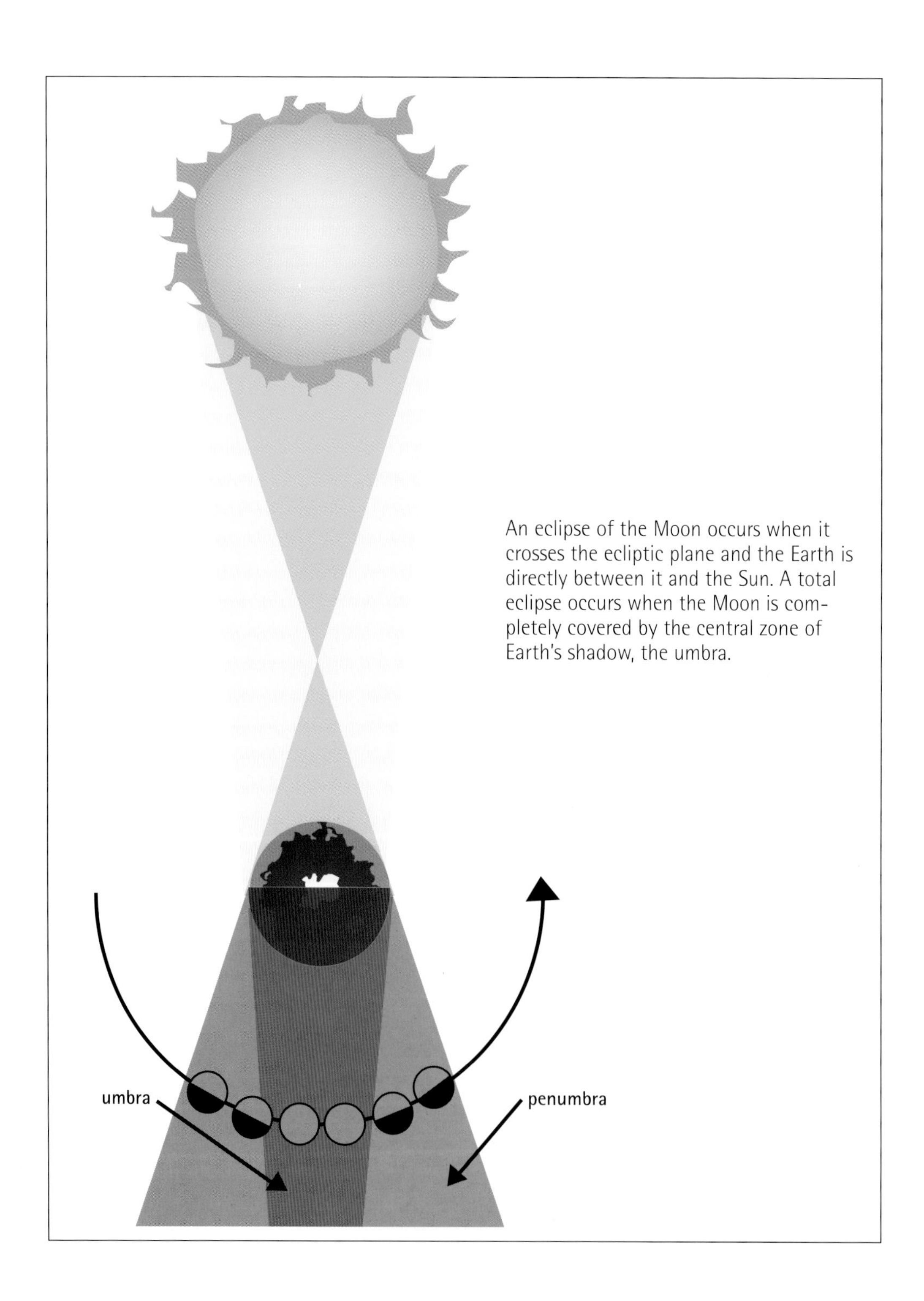


An eclipse of the Moon occurs when it crosses the ecliptic plane and the Earth is directly between it and the Sun. A total eclipse occurs when the Moon is completely covered by the central zone of Earth's shadow, the umbra.

# Mercury

| | Mercury | Earth |
|---|---|---|
| Apparent visual diameter (arc seconds) | 4.9–10 | - |
| Axial tilt (degrees) | 0 | 23.45 |
| Equatorial diameter (km) | 4 879 | 12 756 |
| Surface gravity (metres per second$^2$) | 2.8 | 9.78 |
| Magnitude (max. brightness) | –1.3 | - |
| Mean distance from Sun (km) | 57 910 000 | 149 597 870 |
| Maximum distance from Earth (km) | 221 920 880 | - |
| Minimum distance from Earth (km) | 77 269 900 | - |
| Number of known moons | 0 | 1 |
| Orbital inclination (degrees) | 7.004 | 0 |
| Orbital period (days) | 87.9 | 365.2 |
| Primary atmospheric composition | potassium/sodium | nitrogen/oxygen |
| Rotational period (d/h/m) | 58d15h30m | 0d23h56m |

Moving outward from the immense, searing heat of the Sun we encounter the small, iron-rich world of Mercury. Innermost of all the planets, tiny Mercury orbits the Sun once every 88 days and is perhaps the most heavily cratered body in the Solar System. Second only to Pluto, Mercury's orbit is far more eccentric than all the remaining planets. At closest approach to the Sun (*perihelion*), Mercury speeds by at a distance of 46 000 000 kilometres. This distance extends to a maximum separation of 69 800 000 kilometres at *aphelion*. But Mercury holds yet another second placing to Pluto in terms of its orbital inclination of 7° to the plane of the ecliptic.

This heavily cratered world endures the greatest range of temperatures, from 427°C during early afternoon to –183°C during the night. Since Mercury is the closest planet to the Sun one might reasonably assume it is also the hottest. Venus, however, holds this record, due to planet-wide dense cloud cover and a subsequent greenhouse effect. If you were to view the Sun from the surface of Mercury, it would appear three times the apparent diameter than when seen from Earth and some six times brighter. Rotating once on its axis in 58.6 Earth days, a Mercurian solar day (sunrise to sunrise) actually takes 176 Earth days and is the result of its unusual resonant relationship between its orbital and rotational periods, otherwise referred to as *spin-orbit coupling*. This unusual phenomenon means that Mercury rotates exactly three times on its axis every two orbits around the Sun. From the surface of Mercury during closest approach with the Sun, the Sun would appear to do a little loop along its apparent passage across the Mercurian sky.

This image of Mercury, taken by *Mariner 10* during its 1974 fly-by, revealed the most inner planet to be a heavily cratered world like our own Moon. *(NASA/JPL)*

Recent research has shown that Mercury does possess an ever-so-slender atmosphere with a surface pressure one trillion times less than that of the Earth's. This atmosphere, however, is comprised mainly of hydrogen and helium, possibly derived from the solar wind. Its iron core (roughly 75 per cent of its interior) produces a magnetic field greater than that of Mars. But in comparison with Earth it is substantially weaker, yet strong enough to deflect most of the solar wind by creating a 'bow shock' effect around the planet.

The surface of Mercury resembles that of our own Moon with heavily cratered highlands, scarps and faults, but it differs slightly due to its unique geologic characteristics and location. In 1974 the probe *Mariner 10* revealed Mercury to possess, like the all-too-familiar lunar maria, smooth lava-filled plains—remnants of the final stages of bombardment by early Solar System debris. One of the most striking impact scars is that of the multi-ringed Caloris Basin with a diameter of 1300 kilometres.

After three decades, NASA plans to return to Mercury with the *Messenger* probe by 2005. It will produce a complete global map of the surface, with an emphasis on determining the origin of its high-density make up, and study its interior and magnetic fields while also determining whether or not Mercury possesses any water-based ice deep in its polar craters where temperatures are extremely cold. The mission will involve two Venus fly-bys followed by two initial Mercury fly-bys until orbital manoeuvres place *Messenger* into Mercury's orbit in about 2009. This still leaves plenty of time for useful contributions from ground-based observations.

## Observing Mercury

At such close proximity to our Sun, Mercury is notoriously difficult to observe from Earth. Until the NASA fly-by of *Mariner 10* in March 1974 our best views, through the world's largest telescopes, provided less information than a view of our own Moon with the naked eye. Most easily seen after sunset or before sunrise, this planet can be observed low in the eastern dawn sky (around greatest western elongation) or in the western evening sky (around greatest eastern elongation). Being so close to the glare of the Sun, Mercury's potential brightness is largely subdued, and as it submerges into a yellow-brown atmospheric soup it twinkles like a faint

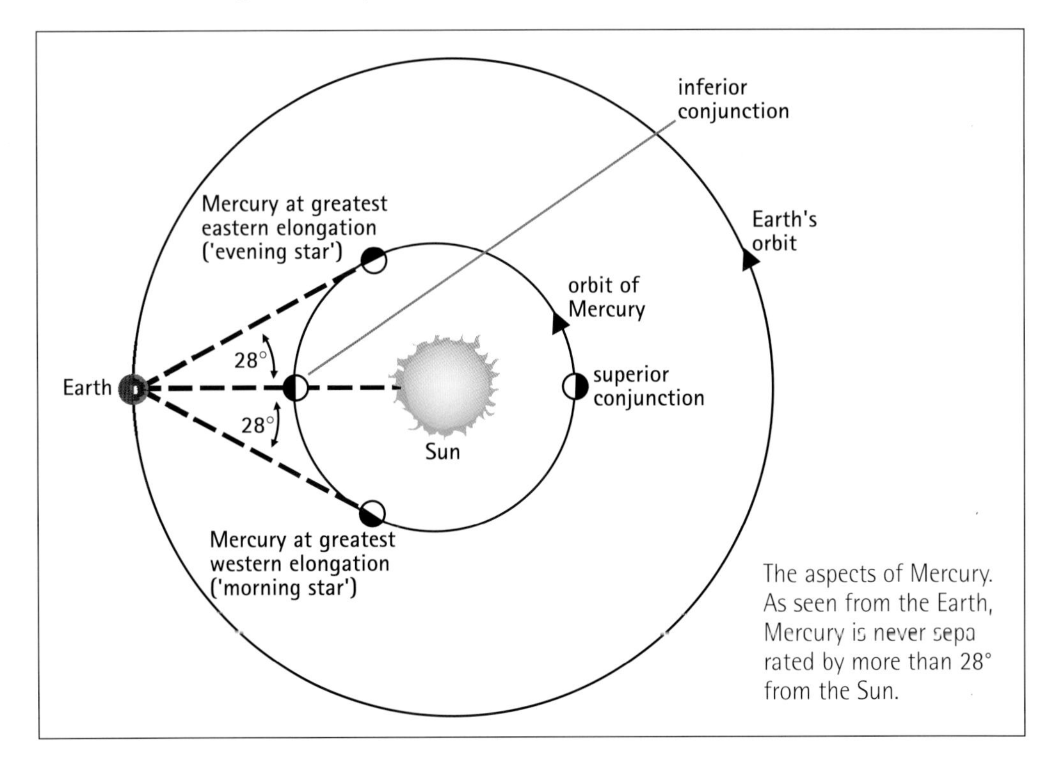


The aspects of Mercury. As seen from the Earth, Mercury is never separated by more than 28° from the Sun.

star, often making it difficult for beginners to locate.

Since Mercury is an inferior planet it is impossible to observe at either superior conjunction or inferior conjunction—the exception to the latter being when it transits the disc of the Sun. In theory, the best time to observe Mercury is when the angle of separation between it and the Sun is at or near its greatest. Depending on the time of year, greatest elongations of Mercury from the Sun range from a minimum of 17° to a maximum of 28°.

Mercury is never far from the Sun, making it a difficult target to observe. The safest time to view the planet is just after sunset when there is no chance of inadvertently catching a blinding glimpse of the Sun. Mercury is seen here sinking into a soupy western evening sky.

It is important to note, however, that greater angular separation does not necessarily translate into improved observing opportunities. At a given time of the year, and based on the observer's location in latitude on the Earth, the angle of the ecliptic plane at dawn or dusk will determine how high Mercury appears above the horizon and for how long it can be observed before it sets. For observers at mid-southern latitudes, the best views are typically around the autumn period (March) when Mercury is in the eastern morning sky, and the late winter to spring months (September) in the western evening sky.

There are two ways in which telescopic observations of the planet can be made: safely after sunset or before sunrise; and during the daytime, which involves both careful planning to locate the planet and extraordinary safety precautions.

For absolute beginners, the first option (preferably after sunset) is a good and safe starting point. When the Sun is below the horizon there is no risk of inadvertently catching a dangerous and potentially blinding glimpse of its intense light while scanning for Mercury with your telescope. Bathed in pre-dawn or post-sunset sky glow, it can be a challenge to find with the naked eye. An ephemeris (table of planetary positions) will serve as a useful reference in order to ascertain the best times for viewing and locating Mercury. However, before dawn or just on dusk you should try to locate Mercury as soon as possible in order to take advantage of its greater elevation above the horizon. A pair of binoculars will greatly improve your chances of locating it while daylight prevails. Once you have located it, take note of its height and direction with respect to a distant landmark then try locating it in your telescope's finder.

## At the eyepiece

Most observers find Mercury to be a difficult and disappointing subject, particularly through a small telescope. Sadly, it is largely neglected by observers due to its telescopically diminutive size and shimmering appearance in the eyepiece. Some are quickly motivated to search for more visually stimulating targets elsewhere. In extremely poor atmospheric conditions it can be quite difficult if not impossible to determine, even roughly, the planet's phase. Behind the bubbling and boiling air layers of the day's heated skies, Mercury appears to warp and twist like some microscopic living cell, as its reflected sunlight is refracted in all directions through the densest parts of our atmosphere. The best time of the day to observe the planet is usually around dawn before the Sun heats up the atmosphere. If you track for some time into mid-morning, while the sky is still slowly warming, you can take advantage of Mercury's higher elevation, increasing your chance of sharper views.

Mercury should not be rejected for its observational difficulty but embraced for the challenges it presents. As seen here on Earth, Mercury exhibits phases throughout an apparition just like those of the Moon. Sometime after superior conjunction Mercury slips gradually higher into the western evening sky. Using a magnification of around 250X in favourable conditions you can begin to observe a tiny gibbous disc. After reaching greatest eastern elongation it then starts drifting back towards the Sun as it approaches inferior conjunction. During this period its increasing angular disc size and crescent phases are more easily noted. After Mercury has passed through inferior conjunction it then appears in the eastern dawn sky. Once it attains enough separation from the Sun to be detected, you can then observe its disc diminishing in size, while passing from crescent to half and gibbous phases. After greatest western elongation it then recedes back towards the glare of the morning Sun once more until it reaches superior conjunction and the whole process starts over.

Along with Mars, Mercury is the only other planet to offer views of its surface, but unlike Mars these moments are fleeting and require diligence, patience and an awareness of what to look for. A picture of Mercury showing evidence of subtle surface markings will certainly generate more interest and publication potential than an average picture of the other, more easily photographed planets.

# Daytime observing

For most newcomers, the idea that some of the planets can be observed during the day comes as somewhat of a surprise. Venus is one such planet and can actually be seen with the naked eye if one knows where to look. Jupiter is also a daytime telescopic target.

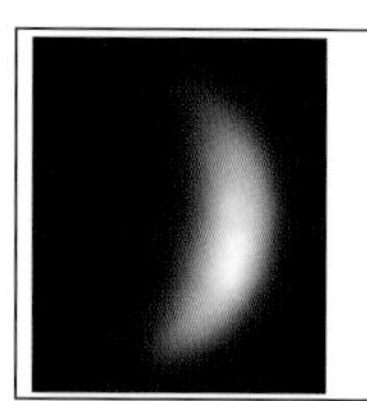

Your best chance of capturing those fleeting albedo features on Mercury is by observing the planet in the early daytime hours before the Sun has heated the atmosphere. The more stable morning air and higher elevations of the planet can result in improved seeing. Although no albedo markings are present in this picture, it is indeed a daytime exposure of Mercury taken with a 250mm (10in) Newtonian.

Daytime observations can be rewarding but require a good knowledge of polar alignment and the use of setting circles on your telescope mount. However, today's more modern digital drive (GOTO) telescopes make the task a whole lot easier for beginners. **Remember, absolute caution must be exercised at all times when observing near the Sun**, as is the case with Mercury.

The main advantage of daytime observations of the inner planets is the ability to view them overhead through thinner layers of the Earth's atmosphere. However, this

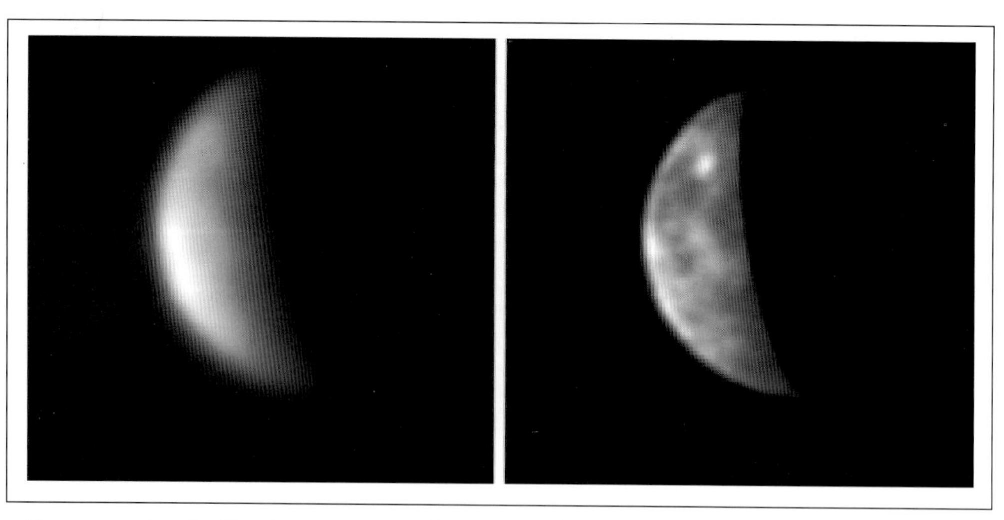

Mercury captured with a low-cost monochrome surveillance camera from the 1.5m (60in) telescope at Mount Wilson on 29 August 1998. At left, thirty averaged frames (to cancel out background 'noise') show only subtle bright and dark markings. At right, a special deconvolution image processing is applied, revealing more detail. *(Courtesy Jeff Baumgardner, Michael Mendillo and Jody Wilson, Boston University)*

is not always as promising as it sounds since the Sun has heated the atmosphere during the course of the day, creating air turbulence that can greatly distort the view. But on rare occasions turbulence can be minimal, offering incredibly sharp and fleeting views of the planet. Another factor to consider is heating of the telescope as the Sun beats down on it while observing. This can introduce unwanted local convection currents within the instrument that will further distort image clarity.

During broad daylight, Mercury is a difficult subject to locate in the eyepiece, especially on your first attempt. The only thing making it easier to locate in later attempts is simply knowing what to expect. Awash in a boiling pale blue sky, Mercury's overall visual contrast suffers greatly due to its proximity to the Sun, diminutive size and low surface brightness. It is easiest to use a low power eyepiece, such as a standard 25mm (1in) one, yielding a wide field of view (FOV) when undertaking your search. At low powers, Mercury appears very small, but when it reaches an angular size around 7 arc seconds its phase is fairly easily noted depending on the seeing. If you have observed Venus in daylight through a telescope before then don't be fooled into believing that Mercury will be as prominent or easily detected. One can stare into the eyepiece for some time before its faint, yellowish half-lit phase becomes obvious.

To assist you in your daytime pursuit of Mercury you will need a current ephemeris for the positions of the planets for an anticipated observing time. With this celestial timetable of coordinates you can select a preferred benchmark such as the Sun, Moon or perhaps Venus with which to calibrate your telescope's setting circles and also for setting the correct focus. Sharp focus is paramount to ensuring you can locate faint Mercury. Unlike night observing, the slightest out-of-focus image makes Mercury virtually impossible to detect. Other suitable pre-focus setting targets include a distant mountain or high patch of cloud in the sky.

Be sure to position your telescope in an area with shaded relief from the Sun. This is worthwhile for three reasons: eye relief from the Sun's glare at the eyepiece; to avoid inadvertent and dangerous encounters with the Sun; and to give relief from the heat during protracted observing during summer months. If no shaded areas are convenient, use a towel or sheet to cover your head and the eyepiece to avoid external reflections from sunlight and other distractions.

A telescope of 130mm (5in) and preferably larger will open the doors of visual opportunity. Some observers have reported brief glimpses of low contrast dusky surface markings on Mercury. In terms of fleeting events, the human eye is indeed a marvellous visual detector. If you wish to record such events, video is a perfect medium. Like our Moon, these regions of varying brightness are Mercury's equivalent to the lunar highlands and dark plains. Italian observer Giovanni Schiaparelli had been studying Mercury since 1881 and was quick to adopt daytime telescopic observations in the hope of sharper views. At the eyepiece of a 218mm (8.6in) refractor, Schiaparelli recorded a network of vague, low albedo regions amidst brighter zones not too dissimilar in appearance to the early Mars drawings by Percival Lowell.

Before observing Mercury, take care to ensure your optical system is clean and free of dust that might present misleading and perhaps illusory artefacts to the image. Whether you choose to make a drawing or utilise video, make every effort to carry out further observations over several days to determine if these vague markings present the same rough appearance, though slightly shifted by the planet's rotation. This will confirm any doubts and reward your effort.

Try using coloured filters such as red Wratten No. 23A and 25 to reduce background sky brightness and improve surface image contrast.

## Transits of Mercury

A transit of Mercury or Venus is indeed an exciting event due to the rare nature of such occurrences. Since Mercury's orbit is inclined to the plane of the Earth's orbit by angles of several degrees, at inferior conjunction it usually appears to pass above or below the Sun. But on rare occasions, when the line-up is just right, it passes directly across the disc of the Sun and appears as a black dot silhouetted against the solar disc. Such phenomena are called *transits*. Mercury transits occur about thirteen or fourteen times during each century and in the months of May or November. During a transit, Mercury's tiny jet-black disc appears even darker than the umbra of a sunspot. The observable angular size of its silhouetted disc as seen against the bright photosphere of the Sun is vastly different from that of a Venus transit. Mercury is physically about two-and-a-half times smaller than Venus and since it is also nearly twice as far away from Earth as Venus is at such times, Mercury presents a disc five times smaller—about 12 arc seconds. Thus comparatively speaking it is more difficult to observe a Mercury transit than a Venus transit. Compared with the angular disc size of the Sun, Mercury is about 1/158 its diameter and is in fact too small to observe without the aid of a telescope. But you certainly don't need a large telescope to view a transit. A small telescope yielding a magnification of 50X to 100X is perhaps the ideal power to use. Since you will indeed be observing the Sun, your telescope must be fitted with a solar aperture filter for safe viewing and photography of the event.

During the unique grazing transit of November 1999 and the more recent 7 May 2003 event, many amateur astronomers managed to successfully record the event using conventional 35mm SLR film cameras, digital and video cameras mounted on small telescopes with solar filters. Some observers prefer to record the Sun by projecting the eyepiece image onto a specially fitted telescope solar projection screen. And, of course, you can always photograph or video the image projected on the solar screen itself.

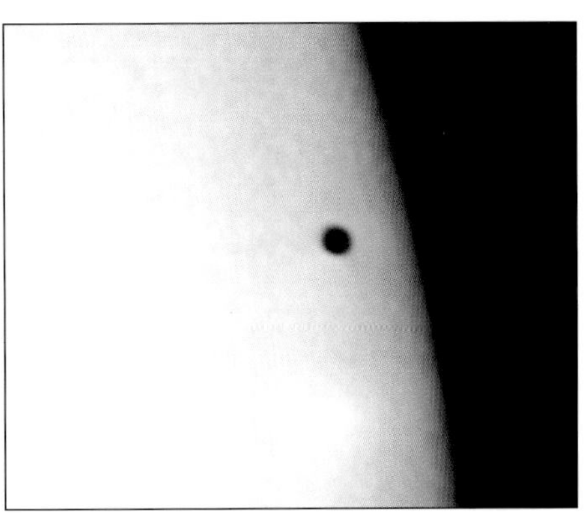

Mercury transiting the Sun, taken on 7 May 2003 from Sydney. After fitting a telescope with a solar aperture filter, this picture was captured using a simple low-cost video camera with two stacked Barlow lenses. Many amateurs around the country used similar tools, including digital cameras, to capture this marvellous and somewhat rare event.

# Ingress and egress

There are four important moments of contact that amateurs can record and which are useful to scientists measuring subtle changes in Earth's orbit and distance. They are the *ingress* and *egress* stages.

## Ingress

- Contact point one—the leading edge of Mercury first encounters the limb of the Sun.
- Contact point two—the trailing limb of Mercury's disc is just within the inner edge of the solar disc before proceeding to cross it.

## Egress

- Contact point three—Mercury's leading edge makes its first contact with the opposing inner edge of the solar disc.
- Contact point four—the final contact, when the planet's trailing edge can last be seen to cause a slight black indentation in the outer limb of the solar disc. This is the end of the transit.

Each point of contact must be recorded in conjunction with an accurate time source. Times must also be accurate to within seconds to be of any real value. An observer can call out the instant of each contact point either onto a tape recorder while using a stopwatch or can simply record the event on video with the video clock date and time superimposed on the image. Of course your video's clock will also need to be set to a reliable reference clock beforehand.

In normal filtered visible light, observing contact points one and four is basically not possible since the area surrounding the solar disc is black, as is the disc of Mercury. If you have access to a hydrogen-alpha filter then the planet may be seen at these contact points, in an incredible spectacle set against a prominence or the inner chromosphere. Just prior to contact point two, the black drop effect can be seen. When this occurs, Mercury appears as though attached to the Sun's limb by a thin black column. Once detached, the planet is completely surrounded by sunlight and this is the true moment of contact point two. Contact point three occurs in exactly the reverse order.

Note that near-actual moments of first and last contact with the solar limb can be quite difficult to observe, especially if the local atmosphere is severely rippling with turbulence.

The next opportunity to observe a Mercury transit occurs in November 2006, so make sure your camera is primed and ready to permanently record this rare event. After all, our next opportunity won't be until 2016.

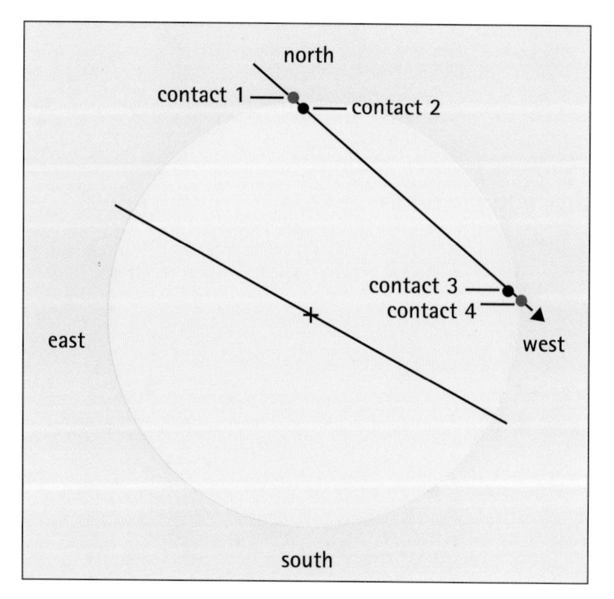


Ingress and egress contact points during the Mercury transit of 2003.

# Venus

| | Venus | Earth |
|---|---|---|
| Apparent visual diameter (arc seconds) | 10.0–64.0 | - |
| Axial tilt (degrees) | 177.36 | 23.45 |
| Equatorial diameter (km) | 12 104 | 12 756 |
| Surface gravity (metres per second$^2$) | 8.87 | 9.78 |
| Magnitude (max. brightness) | –4.4 | - |
| Mean distance from Sun (km) | 108 200 000 | 149 597 870 |
| Maximum distance from Earth (km) | 261 039 880 | - |
| Minimum distance from Earth (km) | 38 150 900 | - |
| Number of known moons | 0 | 1 |
| Orbital inclination (degrees) | 3.394 | 0 |
| Orbital period (days) | 224.7 | 365.2 |
| Primary atmospheric composition | carbon dioxide | nitrogen/oxygen |
| Rotational period (d/h/m) | 243d0h36m | 0d23h56m |

Moving away from the Sun and beyond the orbit of Mercury we find the planet Venus, in an orbit roughly 108 000 000 kilometres from the Sun. Some planetary scientists refer to Venus as Earth's sister planet but in terms of living conditions nothing could be further from the truth. When comparing the two, terrestrially speaking, there are similarities in terms of radius, average density and surface gravity. Even the presence of thick cloud cover had early astronomers thinking that perhaps some tropical, steamy world flourished below. As a place to visit, however, Venus presents an entirely different picture—even for Earthly visitors of the mechanical kind.

This hostile planet rotates in retrograde and, to an observer on the surface, the Sun rises in the

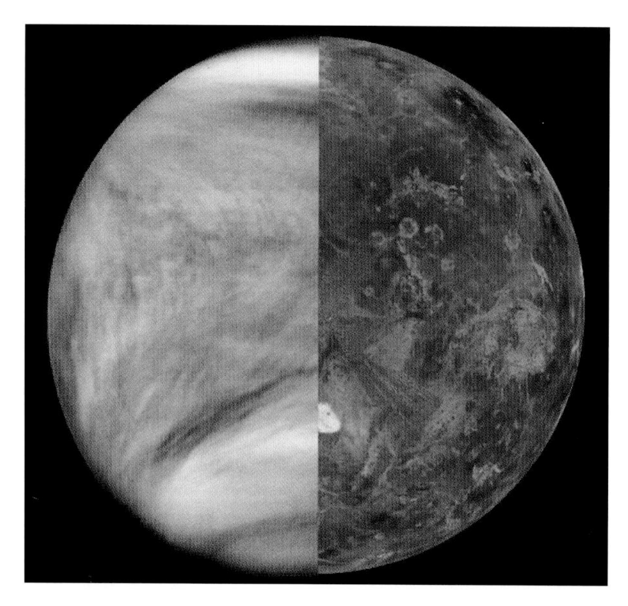

Two views of Venus. The left half shows the clouds that cover the globe of the planet as taken from *Pioneer*. The right half shows the planet's surface as revealed by the *Magellan* spacecraft. *(NASA/JPL. Image compiled by the author)*

west and sets in the east. One day or sidereal rotation of Venus lasts 243 Earth days and is longer than a Venus year, which lasts 225 Earth days. Just thinking about this phenomenon is in itself quite bizarre. How this strange backward rotation occurred still remains a mystery.

Until the 1960s Venus was an enigmatic world, a planet completely covered by a dense yellow-white cloud deck. The highest cloud deck (roughly 70 kilometres above the surface) traverses the planet's globe once every four days, travelling in a cold (–43°C) 360-kilometre per hour jet stream. Descending slowly, the atmosphere grows progressively warmer and more dense until at an altitude of roughly 48 kilometres there is a break in the clouds where sulfuric acid droplets and any solid particles are broken down into water, sulfur dioxide, oxygen and other sulfur-based compounds. At this point, and in a cyclical fashion, these materials are then re-elevated to the higher cloud decks, similar to the thermal characteristics of water-based clouds here on Earth.

At the surface Venus is a hellish, hot world with a poisonous carbon dioxide atmosphere. The planet's global cloud cover makes it a thermal hothouse and fine example of the runaway greenhouse effect. If you stood on its surface and looked out to the horizon it would seem as if you were under water; distant objects would appear to ripple and shimmer, while only the nearest objects, up to 100 or so metres away, would appear with any amount of clarity.

In 1968 the Russian probe *Venera 4* was the first to enter Venus' atmosphere, recording an atmospheric pressure roughly twenty times that on the Earth's surface before failing during descent. Its failure is no surprise—later missions to the planet measured the average surface air pressure at a crushing ninety times that on Earth. This is equivalent to being 1 kilometre below the ocean but in blistering temperatures reaching 480°C, hot enough to melt lead. This runaway greenhouse effect makes Venus even hotter than tiny Mercury, which is some 50 million kilometres closer to the Sun.

After failures to land a craft on earlier missions, because of the extreme atmospheric pressures and mind-blowing temperatures, Russian scientists tested a new generation of robotic landers in highly heated, highly pressurised simulation tanks. These successful probes—*Veneras 9* and *10* in 1975, and *13* and *14* in 1982—briefly transmitted six panorama surface images of small- to medium-sized basaltic rocks before their eventual demise in the hostile conditions. Arriving at Venus in August 1990, and later placed into a stable near-polar orbit, the *Magellan* spacecraft mapped almost 98 per cent of the planet's surface with radar, building up a composite picture of the global topography with unprecedented resolution. Several pictures revealed a completely alien terrain comprising strange spider-like domes, bizarre and massive shield volcanoes and long tracts gouged out by planet-wide lava flows.

It is believed that the planet has undergone global volcanic resurfacing since its formation. What a spectacle that would have been to witness.

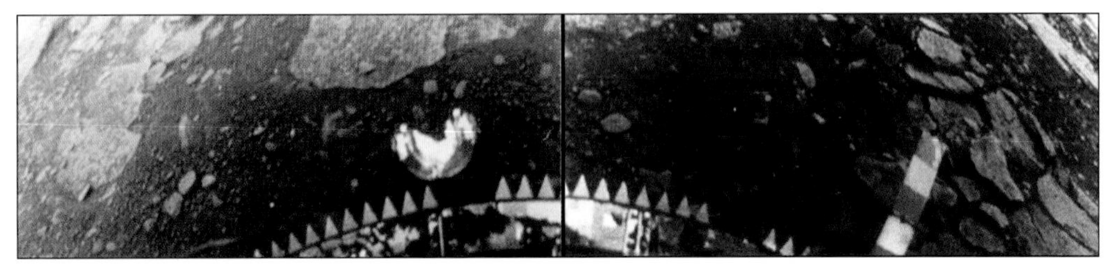

A panorama surface image of Venus taken by the Russian *Venera 13* spacecraft lander on 1 March 1982. The first of the *Venera* missions to include a colour camera, it survived a little over two hours. Flat slabs of rock and soil can be seen with a small peak at the distant horizon in the top right corner. *(NASA and the NSSDC)*

## Observing Venus

Unlike Sun-hugging Mercury, Venus is easy to locate and observe before sunrise or after sunset, but never reaches more than 47° angular separation from the Sun. It is often called the morning or evening star, as is Mercury. In fact, Venus is so bright it can be seen with the naked eye during the daytime for several months of the year.

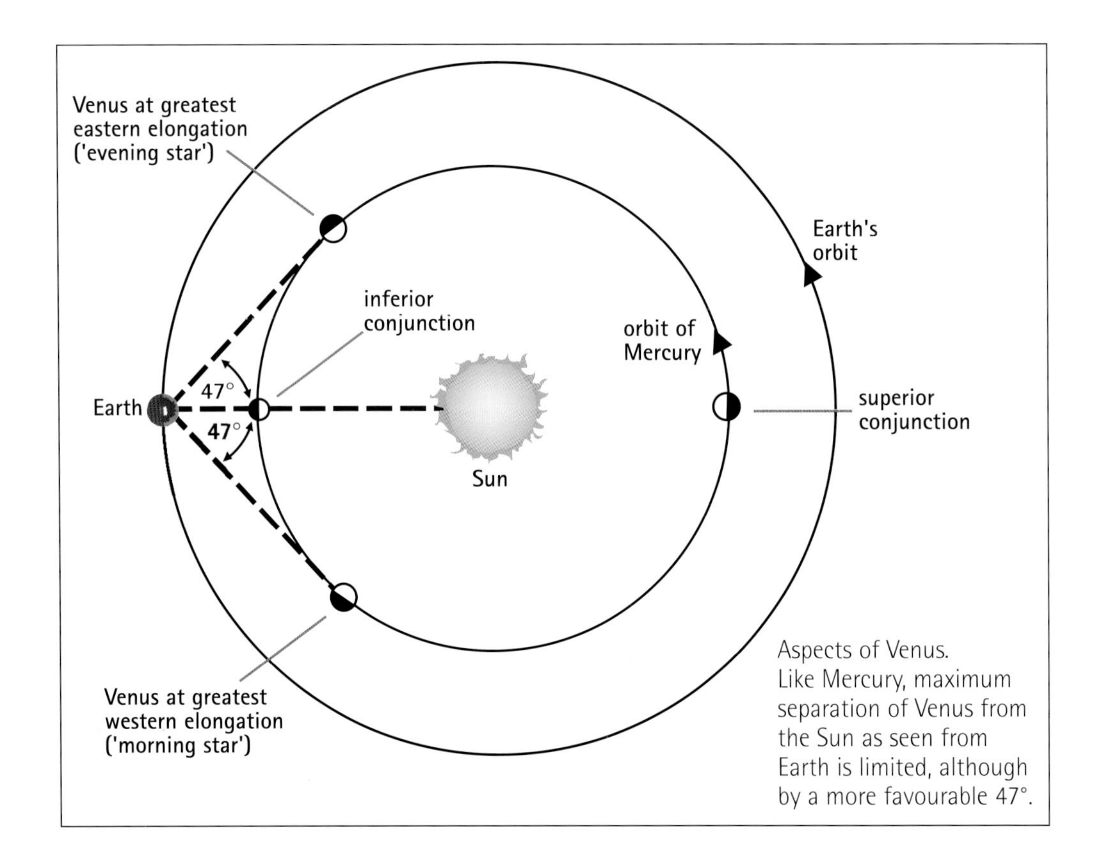


Aspects of Venus.
Like Mercury, maximum separation of Venus from the Sun as seen from Earth is limited, although by a more favourable 47°.

When Galileo observed the planet in 1610, the first evidence of its Moon-like phases became apparent. Its changing phases can be observed through any small telescope, from a near fully lit globe (when the planet is close to superior conjunction) to a hair-thin crescent around inferior conjunction. Venus achieves maximum brilliance shortly after greatest eastern and before greatest western elongation, when its magnitude can reach –4.4. This is bright enough to cast a dim shadow under dark country skies and many claims of UFO sightings are often reported at this time.

At closest approach, a distance of approximately 39 000 000 kilometres separates Venus from the Earth. With a similar equatorial diameter to Earth, such close encounters around inferior conjunction means Venus' angular disc size swells to around 64 arc seconds exceeding even that of Jupiter at opposition. Its large appearance around inferior conjunction even makes it possible to detect the planet's crescent phases through 7 x 50 binoculars. However, in terms of observational satisfaction Venus is sometimes ignored as a target of great interest due to its featureless appearance. If it were not shrouded by such persistent thick cloud, it would most certainly rank higher on the celestial target list. Even though we have now viewed its tortured surface through the global radio mapping of the *Magellan* spacecraft, it is interesting to ponder how it might have appeared had it evolved differently.

At the eyepiece of a telescope, the most striking initial observation is its dazzling brilliance that can almost seem blinding at low powers on first inspection. This is especially the case when its disc is illuminated between 20 and 60 per cent; since Venus' cloud cover reflects nearly 80 per cent of the Sun's light back into space, this is certainly no surprise! If using binoculars, you might consider making two small holes in the centres of the dust caps, thus stopping down the overall aperture and thereby reducing glare and improving contrast between the planet's disc and background sky. Binoculars of 10 x 50 are well suited to the task but should be mounted on a tripod or firm surface. Simply trying to hold them by hand will reveal nothing more than a white blob jumping around erratically in the field of view.

The changing face of Venus captured over several months in 2002. Second from the left, Venus is at dichotomy, while the last three crescent views were taken in broad daylight to allow for the best seeing. At far right, Venus was only 1.26 per cent illuminated and very near the Sun.

At the telescope, using a higher power eyepiece for closer inspection will reduce the amount of glare, but image steadiness is usually the trade-off. Particularly when low in dusky skies, Venus (like Mercury) can take on a shimmering, ghostly appearance. Furthermore, increased magnification will not reveal any more detail so if the seeing is poor then it is best to use a low power eyepiece. Eyepiece filters can greatly improve image contrast. Filters such as a red Wratten No. 23A or 25 can substantially improve edge definition of the limb and terminator.

Once your eye has adapted to the planet's visual brilliance, take a careful look along the terminator. The illuminated portion of Venus often appears dusky along the terminator, even without a filter. If you plan to study the phases of this otherwise featureless planet during the course of an apparition, you might want to make special note of the terminator and cusps. Occasionally one can detect slight deformations along an otherwise smoothly curving terminator. These are very subtle and must be examined with great scrutiny to ensure they are not an illusion, especially if you are feeling tired or atmospheric seeing is poor. These bumps and troughs may be the result of subtle variations in cloud height or brightness, where high cloud patches may be catching the Sun's light like a mountain peak before sunrise. It should be noted, however, that the apparent bulging may be just that—it may in fact only be an optical effect due to a brighter reflective cloud near the terminator. Some observers claim to have seen and subsequently drawn the planet where one of the cusps or horns appears either detached or blunted. Though I have not witnessed this myself there are many credible observations to suggest it does occur.

Around times of inferior conjunction the horns of a crescent Venus can appear notably extended but this is best seen against a reasonably darkened sky. Since Venus is visually very near to the Sun around this time, it is already quite low in the sky at dawn or dusk, thus favourable dark-sky contrast is often compromised by poorer seeing.

Another interesting undertaking has to do with visual perception around the time when Venus is calculated to be at exact half-phase (known as *dichotomy)* as opposed to the time it is observed to be so. Venus' terminator will appear as a perfectly straight line without any curving on either side. This effect is mainly attributed to the diffuse appearance of the terminator in contrast to the much more sharply defined limb, and the difference between observed and calculated dichotomy can vary by up to a few days.

Aside from the vague dusky appearance of the terminator and possible deformations mentioned previously, Venus generally appears featureless at visual wavelengths, but that doesn't mean there is nothing more to see. Dark blue or violet filters such as a Wratten No. 47 can sometimes reveal subtle variations of albedo in the Venusian clouds. These can appear as wispy elongated bands or amorphous patches and are the result of varying wavelengths of scattered and absorbed ultraviolet light high in the planet's atmosphere. Observing and imaging Venus in ultraviolet light can reveal

Using an ultraviolet filter when observing or photographing Venus reveals more than simply a bland white planet. In this video image, subtle banded markings are visible in the upper cloud deck of the planet.

more than just dusky markings at the edge of visual perception. Such filters used with a reflector of 200mm (8in) or greater can greatly improve contrast between the lighter and darker zones. Several amateurs have obtained exceptional results using film and CCD cameras with ultraviolet filters. It is important to note that a reflecting telescope is best suited to ultraviolet imaging due to the highly reflective nature of the aluminium mirror coating at short wavelengths. The multicoated flint glass objective lens in most refracting telescopes acts as a high absorber of ultraviolet light.

# The ashen light

As mentioned earlier (see page 66), a few days before and after a new moon the Moon's night side is bathed in earthshine, the reflected sunlight from the Earth. In the case of Venus, however, having no moon of its own, such an occurrence is impossible. But observers have claimed over the years to have witnessed a sort of ashen light on the night side of Venus and occasionally in the form of a brightening patch seen fleetingly. Such claims are often met with some scepticism, but should not be ruled out entirely. Surely Venus is too far away from Earth, even at closest approach, to reflect the bright light of Earth from its highly reflective cloud tops? The *Galileo, Pioneer* and *Venera* probes recorded electromagnetic impulses during their encounters and some believe these to be the result of lightning activity. Furthermore, recent ground-based observations have, on an extremely limited basis, turned up what could optically represent such lightning effects and maybe only those exhibiting the greatest brightness. With occasional atmospheric cloud thinning in certain places and at certain times, perhaps it is a truly rare but real possibility; general consensus, however, suggests otherwise.

Transits of Venus are extremely rare compared to those of Mercury. During one of his major voyages, in 1769, Captain James Cook and his crew witnessed and recorded the planet's passage across the face of the Sun. This allowed astronomers of the time to make useful calculations in order to determine the Earth's distance from the Sun. The last event occurred in December 1882 and we are extremely fortunate to be living in these times, with an event just around the corner in June 2004. This will be followed by another event in June 2012. No doubt most of us will have passed on before the next occurrence in 2117. Imagine how society will have changed by then!

With even a small telescope and a safe solar filter or h-alpha filter, such an event should not be missed. The large scale of images produced by a CCD video

After first contact, a telescope reveals the disk of Venus silhouetted against the bright limb of the Sun. The background light of the Sun is refracted through the planet's cloud tops, subtly illuminating its trailing limb in contrast to the surrounding blackened sky.

camera will provide stunning views on a television monitor. In fact, you'll be able to record the event with a simple camcorder using maximum optical zoom (although the results will be somewhat less dramatic). When imaging a Venus transit with a telescope, try using the maximum practical power that the seeing will allow. If you want to contribute to a scientific data pool of worldwide observations then you should record the universal times for each point of ingress and egress observed. Rules for observing and recording these stages apply as per the previous details given for a Mercury transit:

## Ingress

- Contact point one—the leading edge of Venus first encounters the limb of the Sun.
- Contact point two—the trailing limb of Venus' disc is just within the inner edge of the solar disc before proceeding to cross it.

## Egress

- Contact point three—Venus' leading edge makes its first contact with the opposing inner edge of the solar disc.
- Contact point four—the final contact when the planet's trailing edge can last be seen to cause a slight black indentation in the curve of the solar disc. This is the end of the transit.

Again, the best way to record the entire event will be with video, allowing playback and closer examination of phenomena such as the black drop effect at a later time. The black drop effect in a Venus transit is somewhat more pronounced than a Mercury event due to the diffusion of sunlight in the planet's atmosphere around the edge of its disc.

The upcoming transits of Venus will be very well documented and present an excellent opportunity for observers around the world to witness and even record a piece of science history, even if they are beginners. During the 2004 transit, Venus will display an angular diameter of around 58 arc seconds. This is huge compared to a transit of Mercury and with suitable eye protection such as solar eclipse sunglasses it will even be visible without the aid of a telescope!

# Mars

| | Mars | Earth |
|---|---|---|
| Apparent visual diameter (arc seconds) | 3.5–25.1 | - |
| Axial tilt (degrees) | 25.19 | 23.45 |
| Equatorial diameter (kms) | 6 794 | 12 756 |
| Surface gravity (metres per second$^2$) | 3.72 | 9.78 |
| Magnitude (brightness) at opposition | max. –2.9 min. –1.0 | - |
| Mean distance from Sun (km) | 227 940 000 | 149 597 870 |
| Maximum distance from Earth (km) | 401 355 980 | - |
| Minimum distance from Earth (km) | 54 510 620 | - |
| Number of known moons | 2 | 1 |
| Orbital inclination (degrees) | 1.85 | 0 |
| Orbital period (years) | 1.88 | 1 |
| Primary atmospheric composition | carbon dioxide | nitrogen/oxygen |
| Rotational Period (d/h/m) | 1d0h37m | 0d23h56m |

With a colourful history of hope and excitement, controversy and science fiction, mysterious Mars, the fourth planet from the Sun, has long held our earthbound attention since ancient times. The outermost of the four terrestrial planets, Mars revolves about the Sun between the orbits of Earth and Jupiter. To date it is the most well-explored planet in our Solar System, and, for many, perhaps the most interesting to observe. For this reason, we'll take a more detailed look in this chapter at the many aspects of the red planet.

## Before the space probe

With the introduction of the telescope in 1609, Galileo Galilei was quick to use this valuable new tool to explore the Moon and the planets beyond the limits of our eyes. One can only imagine the elation he must have felt at the plethora of visual discoveries he was to make. The red planet was one such target to which Galileo soon turned his attentions. Given the crude optical quality of the earliest telescopes, we can only surmise that Mars was among the least visually inspiring subjects. Even through a small telescope of superior optical quality today, Mars can present a visual challenge. Regardless of optical performance, Galileo's acute observations of the planet's tiny disc would enhance our developing knowledge of the true nature of the Solar System. Galileo had not only discovered Jupiter, another world with moons, but also that Venus exhibited phases like our Moon. To top it off, he also saw slight differences in the illumination of the disc of Mars, even though he was hesitant to claim so. These discoveries would lend yet more credence to concept of a heliocentric system.

Super high-resolution image of Mars captured by the Hubble Space Telescope in 2001. *(NASA/STScI/AURA)*

New and improved telescopes—some with awkwardly long focal lengths—were producing sharper views at greater powers. A new era in astronomy had erupted and more observers would come to the fore with substantially more accurate and detailed observations. Christiaan Huygens made an early sketch of Mars showing a dark, wedge-shaped marking on the disc. This marking later came to be known as Syrtis Major but was originally called the Atlantic Canal or Kaiser Sea. Huygens made an estimation of the planet's daily rotation to 24 hours, based on the regular reappearance of this feature. Later observations by Giovanni Cassini refined this period to 24 hours, 40 minutes—not far off the actual period of 24 hours, 37 minutes. Planetary scientists refer to one Martian day as a *sol*. Among other observational achievements to his credit, Cassini also recorded the presence of the Martian polar ice caps.

Several other dark features were later described and mapped. These too were given the names of large bays, seas and oceans, and several of these names are still used today. But why did early observers interpret these features as watery vestiges? One need only look upon the dark features spread across the face of the Moon as seen with the unaided eyes of our ancestors. Before the telescope revealed their true nature, these dark, dry lava plains were called maria (Latin for seas). If for a moment we discard all that we now know about the red planet and place ourselves at the eyepiece of a telescope, it's not hard to understand how this misinterpretation also applied to Mars. Peering into the eyepiece we see a small red-orange disc amidst a barren black sky. As we carefully try to peer through subtle undulations of the Earth's atmosphere, the poles of Mars appear noticeably brighter than the rest of the planet. Darker regions with irregular edges appear to border ochre-hued deserts somewhat comparable to the boundary of a coastline.

This strangely different world possessed more 'apparent' Earth-like characteristics than the other planets. Well-known observers including William Herschel, the discoverer of Uranus, accepted the land-and-sea hypothesis for Mars, but not all were convinced. As the seasons of Mars changed, observers reported a fading and darkening of these features, which took on a more greenish hue at times. A few observers concluded that this might be the result of vegetation tracts fed by water from the melting polar ice caps. This suggestion was further supported in that no-one had ever witnessed evidence of the Sun's reflection off the hypothetical Martian oceans. By the mid- to late-1800s, however, the oceanic idea had gradually begun to wane.

## To see or not to see

Closely studying the red planet in 1877, Giovanni Schiaparelli embarked on the task of systematically mapping features he observed at the eyepiece. During fleeting moments of atmospheric stability, he noted several faint dark lines traversing the planet that sometimes intersected, even broadening at their junctions. While this network of lines lacked the winding routes of many of our vast river systems here on Earth, Schiaparelli was cautious not to speculate as to their possible origin and simply called these wispy markings cannali, meaning channels. Tilted to its orbital plane by 25° and with a day lasting 24.5 hours, Mars seemed tantalisingly similar to Earth. Its polar ice caps changed with Earth-like seasons, while sandy dust storms erupted on its surface and thin clouds moved across its skies; it's little wonder that Mars was considered most likely candidate for harbouring life.

Percival Lowell, a wealthy Boston amateur astronomer, further fuelled this concept and, building on Schiaparelli's observations, perceived and documented more detailed structure to these faint lines. Preferring the term 'canals' he was to speculate on a more intelligent reason for their existence. For whatever reasons, many other observers also claimed to have seen these faint hair-like lines while others could not. Using the 1.5-metre (60in) telescope at Mount Wilson in the United States, George Ellery Hale reported no evidence of canals. Along with other respected observers of the time, Edward Barnard (1857–1923) agreed. We now know that these canals do not exist and that perhaps these observers unconsciously experienced a common trick of the eye that can occur when the brain steps in to create some sort of order from vaguely grouped markings at the edge of our visual perception. In some instances it may have been a simple case of 'I want to believe'.

## Phobos and Deimos

While observers focused efforts on mapping the red planet or attempting to determine its chemistry, others sought to find the elusive moons. After numerous frustrating attempts, an American astronomer, Asaph Hall, announced their discovery in 1877. It's not surprising that observers experienced difficulty in locating the moons since these diminutive Martian worlds orbit very close to the planet and are often lost in its obtrusive glare. Named Phobos and Deimos, these moons are more like mountain-sized boulders, with dimensions of 26 x 18 kilometres and 16 x 10 kilometres respectively.

The tiny moon Deimos is seen here in this four-exposure sequence as it moves along its orbital path around Mars. Phobos and Deimos are both extremely difficult to detect due to their diminutive size and close proximity to Mars—they are lost to its glare as a result. A later image of Mars has been superimposed on the saturated area to show its actual size.

## Advancing techniques

Having exhausted most of what could be visually gleaned from the planet, attempts were made to detect water or oxygen in the atmosphere of Mars using a method called *spectroscopy*.

In short, spectroscopy is a technique utilised by astronomers to analyse and determine the physical and chemical properties of celestial bodies. When continuous white light (like that from the Sun) passes through a prism, it spreads out into its constituent colours from reds to blues as seen in naturally occurring rainbows during a sun shower. When passed through different gases, like those that make up the atmosphere of a planet, the resulting spectrum contains small signatures, characteristic of a particular substance. Spectroscopy was used to discern chemical signatures of the Martian atmosphere as early as 1867 but initially without success.

As both professional interest and telescope time were geared towards more distant targets, systematic planetary observations became the increasing realm of amateurs. Despite the doubt surrounding Lowell's observations of Mars and their lack of acceptance within professional astronomy circles, his dedication and overall contribution was commendable. Committed to a relentless search for a planet beyond Neptune, Clyde Tombaugh found the enigmatic Pluto some years after Lowell's death, from the observatory Lowell had established. In 1905 Lowell published an unconventional idea which utilised the Doppler shift effect to determine the chemical composition of a planet. It was based on the idea that when a planet was near quadrature and approaching or moving away from the Earth at a fast enough rate, the spectral lines for gases of the same composition in each planet's atmosphere would be slightly shifted.

By the mid-1950s astronomers such as C Kiess and William Sinton were taking spectra to determine more about the Martian atmosphere and in particular, any indication of the presence of water. Like several astronomers before them, much of this initial data would prove to be erroneous. One problem had to do with choice of sites for these experiments, where the varying presence of water molecules in our own atmosphere would affect the integrity of the results. Low-resolution photographic plates with slow response times were also inadequate for recording the faint signatures near the infrared end of the spectrum.

In the early 1960s Hyron Spinrad made spectroscopic observations using highly sensitised photographic plates with improved emulsions, and confirmed much higher levels of carbon dioxide in Mars' atmosphere than previously predicted. His results revealed subtle signatures indicating the slightest (if not negligible) presence of water. Spinrad also determined that the planet's atmospheric pressure was much lower than previous calculations and this would have an important impact on the future design of space probes needing to reach the surface safely.

## The *Mariner* revelation

With the advent of modern rocketry, the possibilities for robotic research grew rapidly. A new breed of planetary astronomer, or more fittingly 'planetary geologist', was emerging and interest in Mars was re-energised. When *Mariner 4* flew by Mars in July 1965 its low-resolution cameras beamed back little more than twenty images revealing a gloomy outlook for the possibilities of life. Some researchers were less than prepared for the global coverage of impact craters that soon became evident as pictures came streaming in. Mars appeared to be a depressingly lifeless world and vastly unlike the world HG Wells wrote of in his famous novel *War of the Worlds*. However, this came as no surprise to one Ernest Opik (1893–1985), who'd studied meteors for many years and had predicted with some accuracy the perfusion of impact scarring that might be encountered. *Mariner 4* and the missions that followed abruptly changed our optimistic views. The long debate over frozen lakes and vegetation tracts was finally put to rest.

Mars is an extremely cold and inhospitable planet of rusty rock-strewn deserts, weathered impact craters and a deadly thin ~8 millibar atmosphere made up primarily of carbon dioxide. It is a world where water is unstable and sublimes from ice to gas when heated.

Like orphaned children under a shadow of lowered expectations, we continue to seek out a possible link with a long lost, distant relative. In the years to come we may find indisputable evidence of primordial microorganisms fossilised in the rocks from a sample-and-return mission. But through electronic eyes we are indeed privileged in this lifetime to have witnessed a Martian sunrise and to have gazed across its barren landscapes to distant horizons.

Since the *Mariner* encounters, numerous other spacecraft have travelled to Mars, mapping the globe, measuring the atmosphere, sampling the soil at the surface and beaming back unprecedented high-resolution pictures of a still-mysterious yet more familiar world. Much of the research being done today indicates that Mars is still an exciting planet that may yet reveal signs that we were indeed at one point not alone. At the time of writing, the *Mars Express* orbiter and *Beagle 2*

This spectacular view from the surface of Mars was taken during the Mars *Pathfinder* mission in 1997. Note the dusky coloured sky and the rusty coloured rocks and soil which give Mars its almost reddish appearance when observed from Earth. Viewing the surface of another world even through the eyes of robotic cameras imparts a sense of greater realism when observing the planet at the eyepiece. *(NASA/JPL)*

lander are en route for a new encounter with the red planet. Several other spacecraft are also planned for the next ten years, leading up to an eventual sample-and-return mission.

## Observing Mars

Of the five naked-eye wanderers known in early times, Mars occasionally rivals Jupiter's visual brilliance, achieving a magnitude of –2.9 during favourable oppositions. Its distinctive reddish hue was commonly associated with blood and conflict by the ancients, and this subsequently led to its mythological naming after the Roman god of war. The planet's ruddy hue is due to oxidised soil at the surface much like the brown-red colour in rusty metal. When low to the horizon,

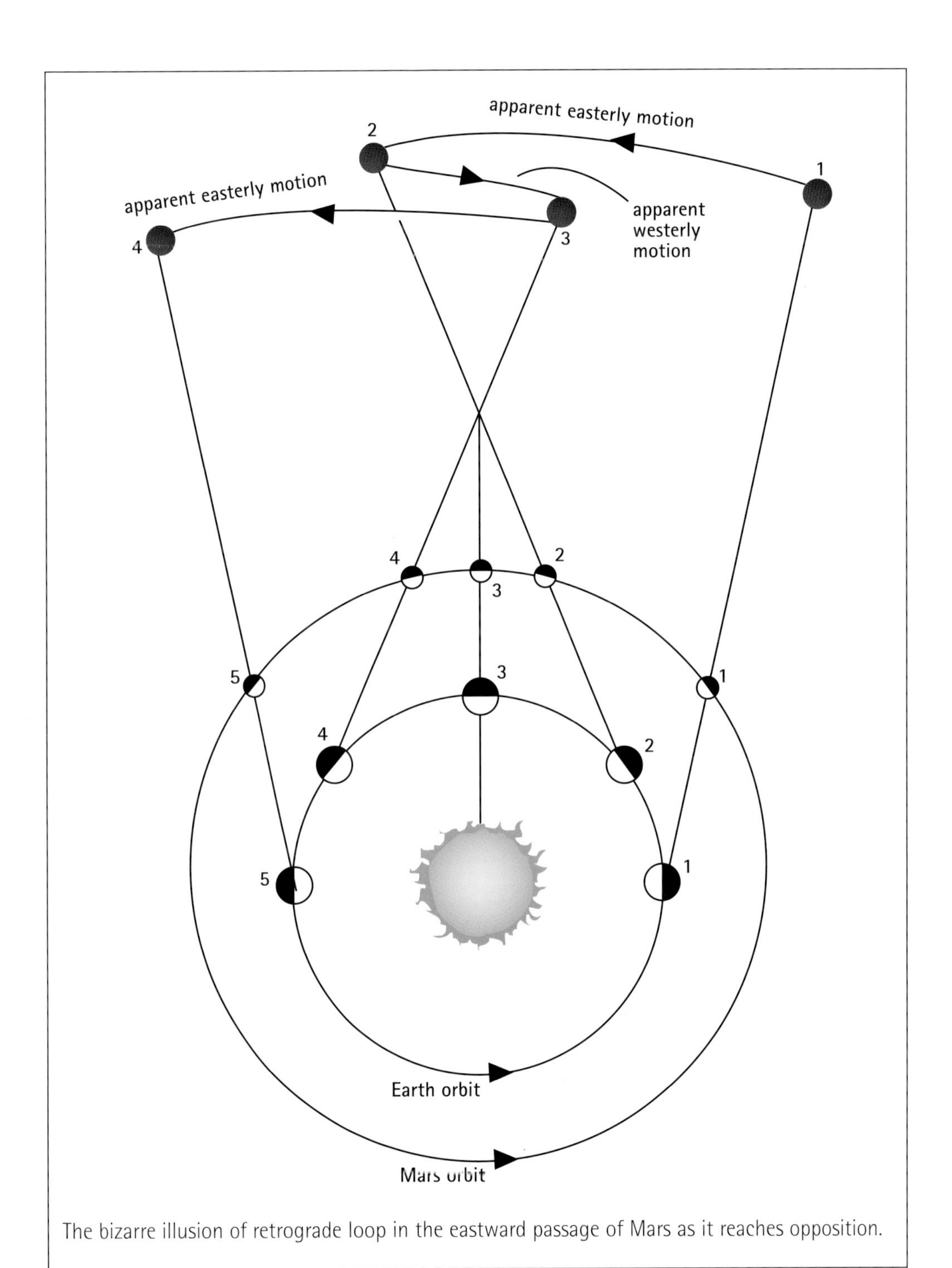


The bizarre illusion of retrograde loop in the eastward passage of Mars as it reaches opposition.

its light is scattered through thicker layers of air and dust particles in our atmosphere, thus giving it a deeper red tone. But as it creeps higher into the night sky it appears a more golden orange colour.

Mars was known by several other names within different cultures. In keeping with a hero god of conflict, it was Ares to the early Greeks and, derived from this name, areography is the study of Mars. To early Egyptians, Mars was known as the Red One or 'the one who travels backward', the latter being a clear indication they had well and truly noticed its apparent retrograde motion around times of opposition. This retrograde or apparent backward motion is an illusory effect as seen from our perspective here on Earth. Going about their individual orbits of the Sun, all the planets appear to drift slowly in an eastward motion through the constellations of the zodiac. Since Mars is the nearest of the outer planets to Earth, this behaviour relative to the fixed background stars is much more obvious. As Earth's orbit catches up to Mars, the red planet's easterly drift appears to slow, then stop. It then drifts westward for a few days, only to stop again then continue along its eastward course. If you were to compile a series of photographs over this period it would reveal a distinct loop in the planet's path among the stars.

Mars is the only terrestrial body to offer more than subtle hints of surface detail observable from ground-based telescopes. When at its furthest point from

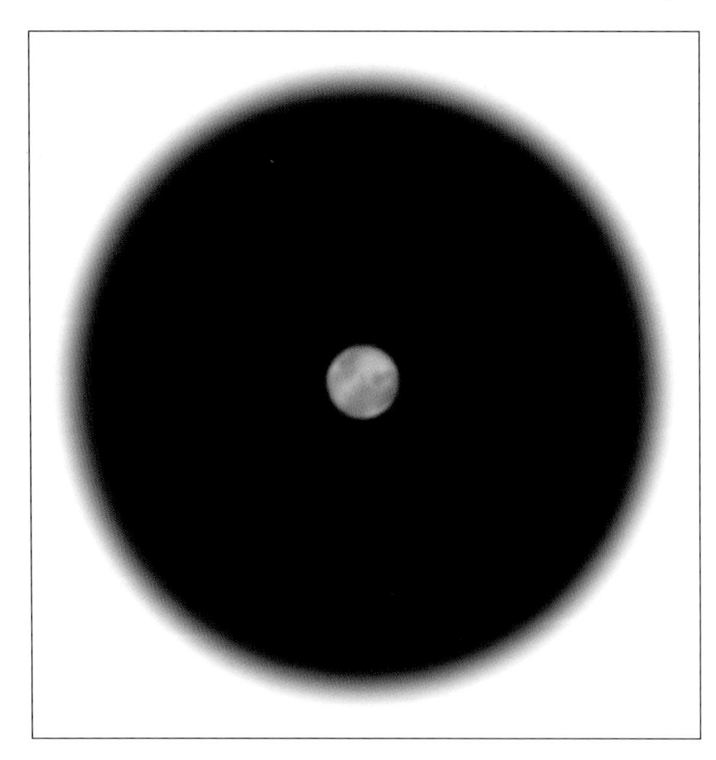

A typical view of Mars at the eyepiece around times of opposition. The seeing must be extremely good for using the highest practical magnifications, making it possible to adequately discern surface features. In this image Mare Acidalium can be seen in the north, along with Mare Erythraeum and Sinus Sabaeus. Bluish limb haze can be seen in the northwest along with the whiter southern polar cap.

This video image was taken with a 250mm (10in) reflector on 4 March 1997 when Mars was at only 13.6 arc seconds. No filters were used and the image is comprised of several stacked pictures from a VHS-quality tape. The north polar ice cap is tilted by nearly 23° toward Earth and Syrtis Major is the dark feature seen on the planet's disc, near the central meridian.

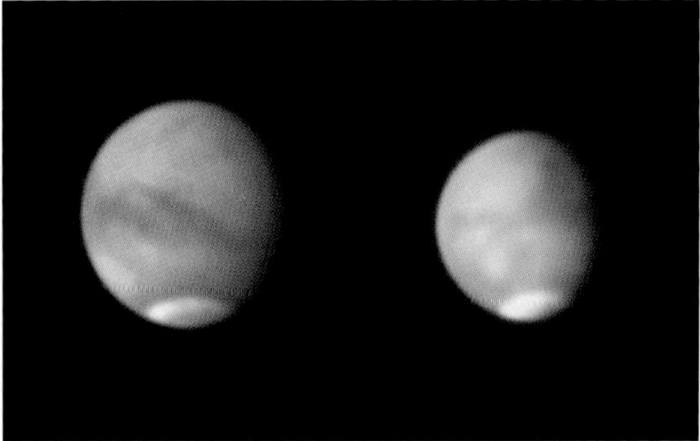

Of the outer planets as seen from Earth, none show as much phase effect as Mars. These pictures were taken almost 2 months before Mars reached opposition in 2003. They show around 90 per cent of its day side. Along with the prominent south polar cap seen in both views, Cimmerium is the dark feature at left while the Syrtis Major region in the view at right is about to move into the night side of Mars. The tri-colour images were captured using an inexpensive CCD video camera and a 200mm (8in) Newtonian telescope. North is up.

Earth (nearer to the Sun in our sky) its apparent angular disc size is about 3.5 arc seconds across and detecting any surface features is a huge challenge to say the least. The best time to observe the planet is when the gap between it and the Earth is narrowest during each particular apparition. These biennial oppositions occur once every 26 months but with each encounter, the distance between Mars and the Earth is not always the same. Because of its markedly elliptical orbit, closest approach (perihelic oppositions) brings Mars to within 55 million kilometres of Earth. At these times, the planet's apparent angular disc size is greater than 25 arc seconds. At least favourable (aphelic) opposition, its distance from Earth is 99 million kilometres, sporting an angular size of only 14 arc seconds by comparison. Like the most recent closest approach in August 2003, such encounters occur at 15- to 17-year intervals. It is worth noting that predicted dates of closest approach and opposition between Earth and another planet do not necessarily coincide. In the case of Mars these dates may vary by up to eight days due to slight differences in each planet's orbital plane.

We don't have to wait 15 years or so to enjoy observing Mars. Even when it presents a small disc of 14 arc seconds, Mars can still reveal a wealth of detail, offering an exciting visual and photographic challenge. It should be remembered, however, that even during favourable oppositions, expectations of a gloriously large planetary globe are unrealistic. Its disc is still relatively small in a telescope and requires higher powers in the order of 2X per millimetre of aperture to clearly discern the many features it can present. Of course, you don't need a huge telescope to say, 'I've seen evidence of dark markings and a bright polar cap on Mars!'. In fact, a typical 60 x 700 beginner's telescope is certainly capable of doing this when the planet presents an apparent angular size of 20 arc seconds or more. But if you want to see more than just subtle markings you'll need a telescope of 100mm (3.9in) or greater to reveal more distinct detail. At certain times in very good seeing conditions, excellent views of Mars can be achieved with an economical 110mm (4.5in) reflector but you will find detail is limited to the largest and/or brightest features. A 150mm (6in) reflector is recommended as the minimum aperture in an affordable telescope for serious detailed views. Reflectors with apertures of 200mm (8in) or greater present more adequately detailed views.

Witnessing the disc of another world suspended in the void of space is always something you won't forget quickly and Mars is no exception. If the air is steady, a telescope with a large aperture will have you gazing into the eyepiece for many hours. If you are not familiar with the planet and its features you should start observing a couple of months before its predicted opposition date. This will help you to become familiar with features and improve your visual acuity for detecting clouds, limb haze, fog patches and surface markings. You can also use this time to experiment with the use of different eyepiece filters to see how they affect what you see.

When the planet is at western quadrature, it displays a phase effect like that of a gibbous moon and can appear almost egg-shaped, an effect accentuated by its polar caps. This can further be seen after the planet has passed through opposition and is once again at right angles to the Sun and Earth at eastern quadrature. At this time, the night side of Mars is present on the opposite limb.

## Surface markings

Just as the four corners of Earth have been charted with a network of latitude and longitude lines, so too have the terrestrial planets and many of the rocky moons of the Solar System. This makes observing Mars even more exciting since surface features observed can be identified on a globe or map. With careful observations you can even create your own map!

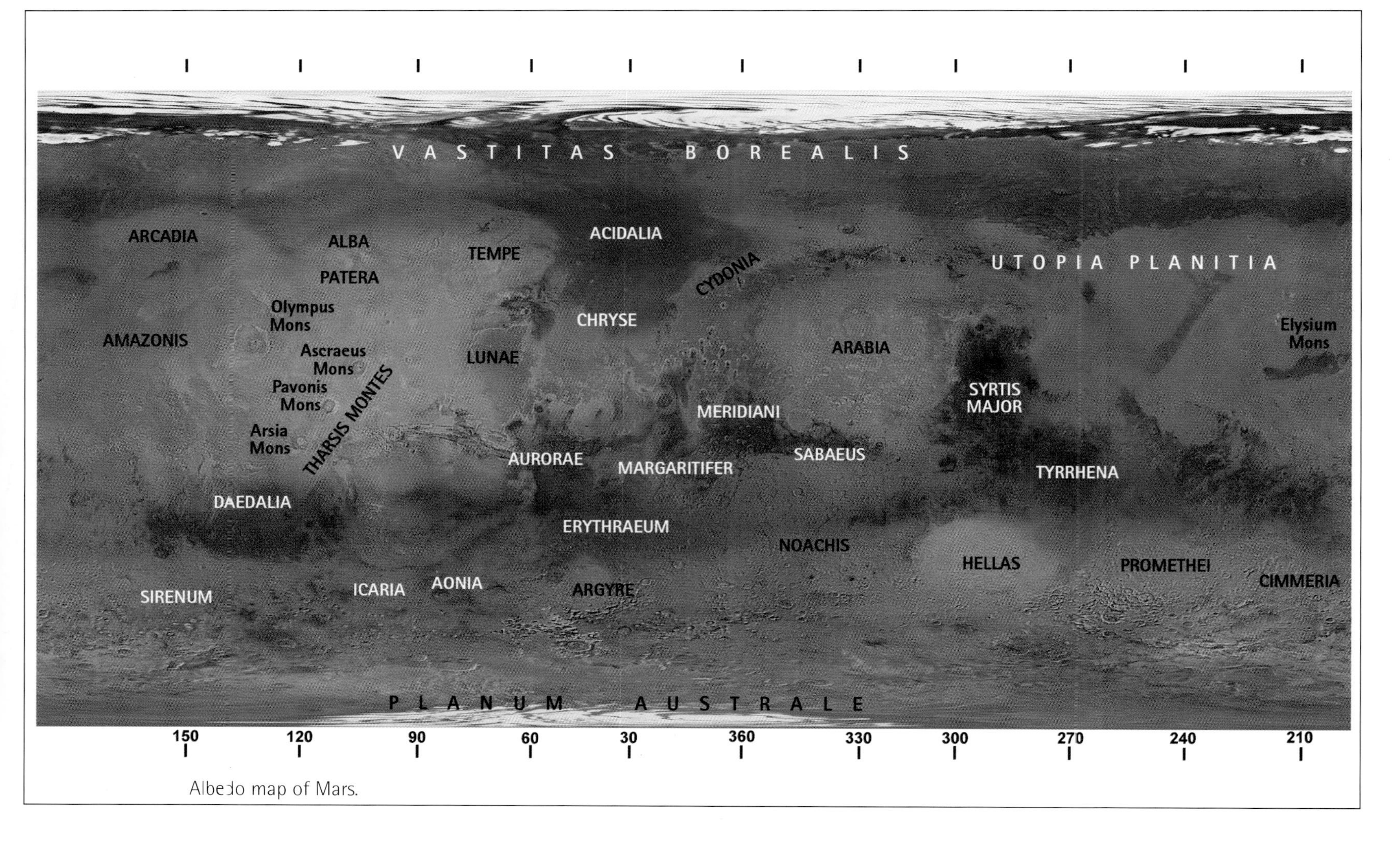


Albedo map of Mars.

Since Mars has a day 37 minutes longer than the Earth, surface features will appear to cross its central meridian 37 minutes later each night. The central meridian is an imaginary line that runs down the centre of a planet's visible disc from one pole to the other. If you were to observe Mars at the same time each night, the planet's surface features will be displaced by 9.5° of longitude compared with the observations from the previous night. If you were to take a snapshot of Mars at this time each night it would take almost 36 terrestrial days to complete a record of the central meridian features while the planet appears to rotate in retrograde. This apparent retrograde rotation is of course an illusion. This is an interesting exercise you might consider trying. Once you have captured enough images or photographs you can try constructing a short animation of the planet's rotation showing its features as they move across its disc

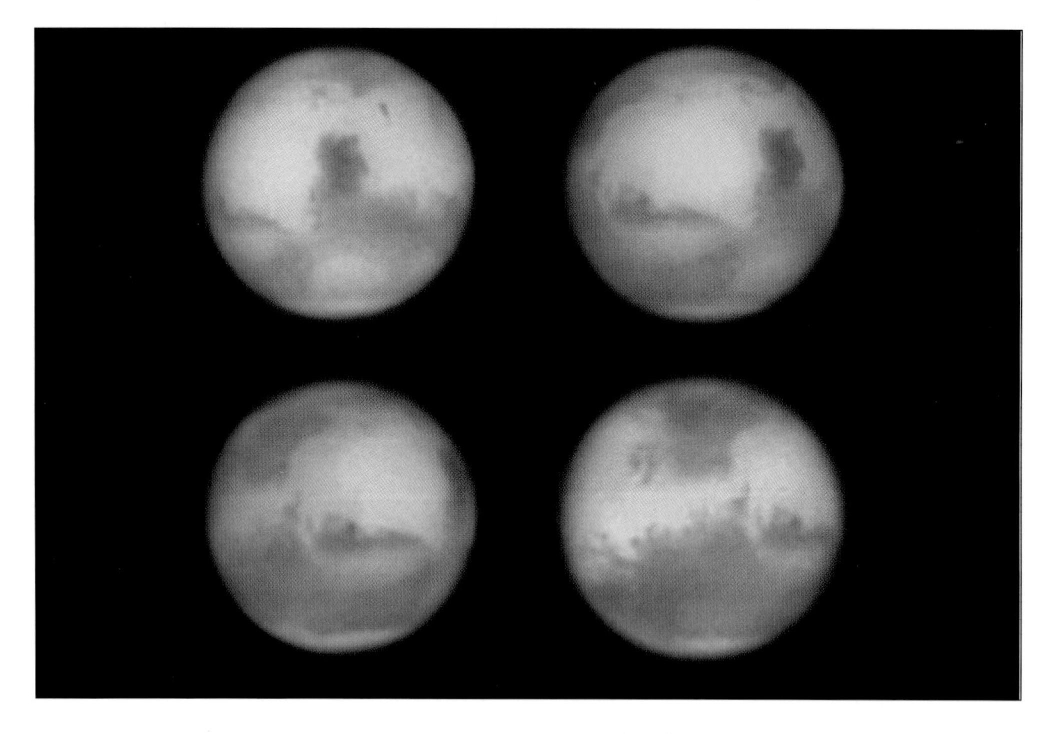

This sequence of Mars was taken in June 2001 using a 0.6m (24in) telescope at Siding Spring in New South Wales. It shows roughly one third of the planet's rotation. At top, Syrtis Major is the prominent feature seen moving across the planet's disc while the lighter, oval-shaped region below it is the Hellas basin. At bottom left, Sinus Sabaeus extends westward to Sinus Meridiani looking somewhat like a hand beckoning someone. At bottom right, Mare Acidalium is the dark region in the Northern Hemisphere while the large craggy region to the south is Mare Erythraeum. North is up.

or a composite picture of each night's observations with the date and time below each image.

In contrast with the lighter surrounding ochre-coloured areas are darker, low albedo regions. Like the vast basalt plains of the Moon, they are also believed to be volcanic rock-strewn areas that are largely free of layered dust cover. At the eyepiece they appear as various hues and contrasts of grey. Over the years amateur observers have recorded their changing shapes resulting from new or removed deposits of dust that have been blown across the planet during large storms.

Red filters Wratten No. 25 or 23A or an orange filter like a Wratten No. 21 markedly increase the contrast between Mars' ochre regions and these dusky, low albedo features. Remember that early observers mistook these dark areas for seas. Take the time to peruse these features carefully and ask—would you have come to this conclusion?

Of these vague dark regions, several of the most prominent can be identified on a modern albedo map of Mars. While the nomenclature for these albedo features originally consisted of the names of classical observers of the time in relation to seas, oceans or lakes, the International Astronomical Union has, since 1919, been responsible for naming newly discovered features. Some features as photographed by space probes have been renamed in keeping with a Latin description of their true geological nature.

At a telescope's eyepice, some of the most prominent albedo features include the regions of Acidalia, Syrtis Major, Meridiani, Noachis, Sabaeus, Sirenum, Tyrrhena, Solis Lacus and Utopia. The most distinct is Syrtis Major, located at 290° longitude. In a 'south is up' view, this region looks something like the continent of Africa.

Of the brighter ochre-hued areas, including the vast Tharsis and Elysium regions, perhaps the most prominent in terms of size, form and contrast lies just below Syrtis Major. Here we see a large circular-shaped feature known as the Hellas basin which exhibits noticeable foreshortening when the planet's north pole is titled towards Earth. Hellas is occasionally covered by a hood of haze during the Martian late autumn and winter periods in its Southern Hemisphere. During southern spring months, frost patches can sometimes be seen in the southern-most parts of the basin which are often associated with nearby cloud activity.

Another bright though much smaller region is called Edom, which greatly contrasts the darker Sinus Sabaeus and Meridiani areas. It is in fact a carved-out area where the large crater called Schiaparelli resides.

## Polar ice caps

Mars has seasonal variability throughout its year, just like Earth, due to its 25.2° axial tilt. However, the length of its seasons is much greater than those on Earth because of Mars' longer orbital period of 687 days. The length of the Martian seasons varies by up to 51 days as a result of its highly elliptical orbit. The polar ice caps are one of the famous Earth-like features of this planet, visible through small telescopes, and both can be seen depending on which hemisphere is tipped earthward. They are comprised of dry frozen carbon dioxide with a thin layer of trapped water-ice below. The polar caps can be seen to grow and shrink alternately at each pole with every Martian season. Since they are so highly reflective in contrast with the rest of the disc, they are perhaps the most obvious feature you will encounter on first inspection. Huge fields of dark sand dunes surround the north polar cap, adding to its overall contrast and distinction. This is especially obvious during the spring and summer months in the north when the caps are thawing, leaving only a button-like tip. A red filter will greatly improve boundary contrast.

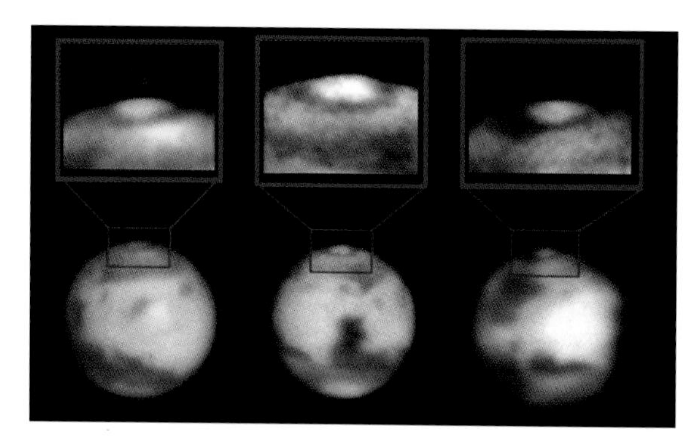

As temperatures warm in the Northern Hemisphere of Mars, the north polar ice cap slowly recedes, leaving a button-like tip surrounded by a dark collar. The collar is known as Mare Boreum or Vastitas Borealis, which is a vast region of enormous sand dunes that encircle the north polar cap. During northern Mars winter, the cap can grow to almost 70°N, forming a large haze over the entire polar hood.

## Clouds, frost and dust storms

Mars may technically be a dead world compared to Earth but its meteorological activity is alive and kicking! Once you begin to hone your observing skills, the realisation that you are actually observing live weather changes on another world from your own backyard can sometimes be overwhelming.

Thin bands of equatorial cloud can occasionally be seen extending across vast regions of the planet's disc. Most obvious are the blue-white clouds on the planet's limbs, sometimes called *limb arcs*. They appear much brighter because they are gathered more densely near the planet's limb as seen from our perspective, and scattered sunlight is reflected with greater intensity. With careful scrutiny and the aid of a blue eyepiece filter such as a Wratten No. 38A or 80A, clouds and limb arcs will be accentuated. A violet filter such as a Wratten No. 47 will brighten high altitude clouds. Other areas where you will often find discrete white clouds during the Martian spring and summer months (most prominent around noon) are Elysium Mons, Tharsis Tholus and Olympus Mons. These are huge dormant volcanoes and clouds gather at their peaks. Olympus Mons is itself the highest known terrestrial feature in the Solar System.

Mars is roughly 1.4 times farther from the Sun than Earth and is subsequently much colder. During a night on Mars, the temperatures drop rapidly and atmospheric carbon dioxide freezes at around –124°C, forming surface frosts and foggy patches close to the surface. In my experience these can be difficult to distinguish from bright clouds; however, unlike poorly defined cloud boundaries, frost patches exhibit more distinct perimeters and move with the planet's rotation. To confirm their true nature, first try a blue filter to accentuate clouds high in the Martian atmosphere. Next, compare its observed brightness to that seen through a yellow or green filter such as a Wratten No. 56. If it appears brighter and more defined in green or yellow light, yet more difficult to observe in blue light, then it is likely to be ground frost.

Dust storms and their development within the Martian atmosphere are of great interest to amateur and professional astronomers alike. Amateur Mars observations were taken into account by NASA/JPL mission scientists during the *Pathfinder* spacecraft encounter with the planet in 1996. This was indeed a great cooperative effort in modern times whereby amateurs

A huge cyclone stirs in the atmosphere near the north pole of Mars in this 1999 Hubble Space Telescope image. This temporary phenomenon was short-lived and measured 1760 kilometres across, even larger than the planet's residual polar ice cap. The eye of the storm is roughly 320 kilometres across. Such large phenomena are visible with simple backyard telescopes. *(Courtesy Jim Bell (Cornell University) and STScI)*

could submit their images to a dedicated international website called *Mars Watch* for review. If there was any indication of a dust storm developing or under way then mission scientists may have been required to re-evaluate the risks and possibly reprogram the craft for the safest descent into the planet's atmosphere.

During the apparition of Mars in April 1999, a large cyclone was imaged by the Hubble Space Telescope (HST) near the northern polar ice cap. Along its greatest axis the cyclone's extent was nearly 2000 kilometres. In late June 2001 a small dust storm that had developed in the Hellas basin region eventually grew to global proportions, obscuring the entire planet's visible surface features for several months. Large dust storms appear brightened with a yellow filter such as a Wratten No. 12 or 15 but exhibit greater boundary definition when using red or orange filters.

**Future dates of opposition**
7 November 2005
24 December 2007
29 January 2010

# Jupiter

| | Jupiter | Earth |
|---|---|---|
| Apparent visual diameter (arc seconds) | 30.5–50.1 | - |
| Axial tilt (degrees) | 3.13 | 23.45 |
| Equatorial diameter (km) | 142 984 | 12 756 |
| Surface gravity (metres per second$^2$) | 24.5 | 9.78 |
| Magnitude (brightness) at opposition | max. –2.9 min. –2.0 | - |
| Mean distance from Sun (km) | 778 330 000 | 149 597 870 |
| Maximum distance from Earth (km) | 968 460 580 | - |
| Minimum distance from Earth (km) | 588 404 520 | - |
| Number of known moons | 61 | 1 |
| Orbital inclination (degrees) | 1.308 | 0 |
| Orbital period (years) | 11.8 | 1 |
| Primary atmospheric composition | hydrogen/helium | nitrogen/oxygen |
| Rotational period (d/h/m) | 0d9h55m | 0d23h56m |

Beyond the orbit of Mars and the sparse rocky plane of the asteroid belt we enter the realm of the gas giant worlds. The first of these planets we encounter is also the largest in the Solar System. Orbiting at a mean distance of 778.3 million kilometres from the Sun, Jupiter is the king of the planets, with an equatorial diameter eleven times that of the Earth—about 142 900 kilometres across. With the exception of the Sun, this massive gaseous world accounts for more than two-thirds of the material that makes up the other bodies of the Solar System; it would take nearly 1300 Earths to fill its volume. The Jupiter system (more commonly called the Jovian system) is a virtual solar system in miniature. At the time of writing the known satellites of Jupiter total sixty-one, and while most of these orbiting worlds are small, asteroid-sized objects, the four largest—Io, Europa, Ganymede and

*Cassini* image of Jupiter captured in 2000. *(NASA/JPL)*

Callisto—can be seen with binoculars from Earth. Galileo was the first to publish the discovery of these moons and named them the Medicean Stars in recognition of his financial sponsors of the time. However, they were later renamed the Galilean moons in his honour.

Jupiter has no solid surface that one could in fact stand on but if it did we would find ourselves crushed beyond recognition. The planet is a huge ball of molecular gas composed of about 81 per cent hydrogen and 17 per cent helium, while the remaining constituents include methane, water vapour and ammonia among other elements. Space probe missions to the planet found that Jupiter generates more than twice the heat it receives from the Sun. It is suspected to have a highly compressed liquid metallic core that acts like a huge dynamo, driving its high-speed winds and huge magnetic field which extends millions of kilometres into space.

Since *Pioneer 10* first flew by Jupiter in 1973, returning our first close-up views of this mysterious world, the planet has since encountered other earthly visitors of the mechanical kind. In 1979 the *Voyager 1* probe measured wind speeds and turbulent storms in Jupiter's atmosphere and discovered the presence of lightning in the planet's cloud tops. During this fly-by it also imaged a faint set of rings that encircle the planet and returned amazingly detailed images of swirling and eruptive clouds, giant circular storms and close-up views of the major moons. Arriving at Jupiter in 1995, the very successful *Galileo* probe imaged the planet and its moons with unprecedented clarity, often achieving one hundred times the resolution of previous missions. Unlike its predecessors, *Galileo* was placed into an orbit around Jupiter where it could undertake protracted studies of the atmosphere and moons. Attached to the spacecraft was a small atmospheric probe, which would later descend into the Jovian atmosphere and take measurements of temperature, chemical composition, pressures, sunlight and energy internal to the planet, cloud characteristics, and lightning. The atmospheric probe survived 59 minutes as it penetrated the violent, soupy Jovian atmosphere to a depth of around 200 kilometres before being crushed by the immense pressure.

Orbiting the Sun every 11.8 years, Jupiter is sometimes referred to as the vacuum cleaner of the Solar System. Due to the strong gravitational influence of this massive planet, the highly elliptical orbits of stray comets or asteroids are sometimes perturbed, eventually sending them to their demise. This was the case in 1994 when for the first time we witnessed the destruction of a comet. Comet Shoemaker-Levy 9 was ripped apart by the enormous gravity of Jupiter during a final close encounter. On their return journey, these fragments were drawn in to Jupiter. Plummeting through the Jovian atmosphere, each icy missile ejected large plumes of material high above the planet's cloud tops. The post-impact scars appeared as a series of dark, Earth-sized spots spread out across the Jovian atmosphere which eventually faded away. Such a collision with the Earth would be devastating and could potentially erase humankind from the face of the Earth.

## Observing Jupiter

Jupiter can be found shining a brilliant white quite near to the ecliptic. Achieving a close second at maximum angular size and brightness to Venus, Jupiter offers a wealth of detail for any backyard observer through most small telescopes. Even 7 x 50 binoculars will reveal its disc and the dance of its four major moons. If all you own are binoculars then you can still observe the nightly changing positions of the moons and make simple drawings in a diary or observer's log.

Fine detailed views of the planet require a telescope of 150mm (6in) or more, but smaller instruments will certainly reveal the wider equatorial belts and zones. When using low powers of around 45X magnification, the first visual impact is perhaps Jupiter's overall brilliance and relatively large size. Of the gas giants, Jupiter is the closest to Earth and thus its proximity and massive diameter account for its large visual appearance. Jupiter's equatorial angular size ranges from 30 arc seconds at its farthest to 50 arc seconds around opposition, while its visual magnitude ranges from –2.0 to –2.9.

Examining Jupiter's disc, we see it has a flattened appearance at the poles and a broadening at the equator, known as a planet's *oblateness*. The amount of

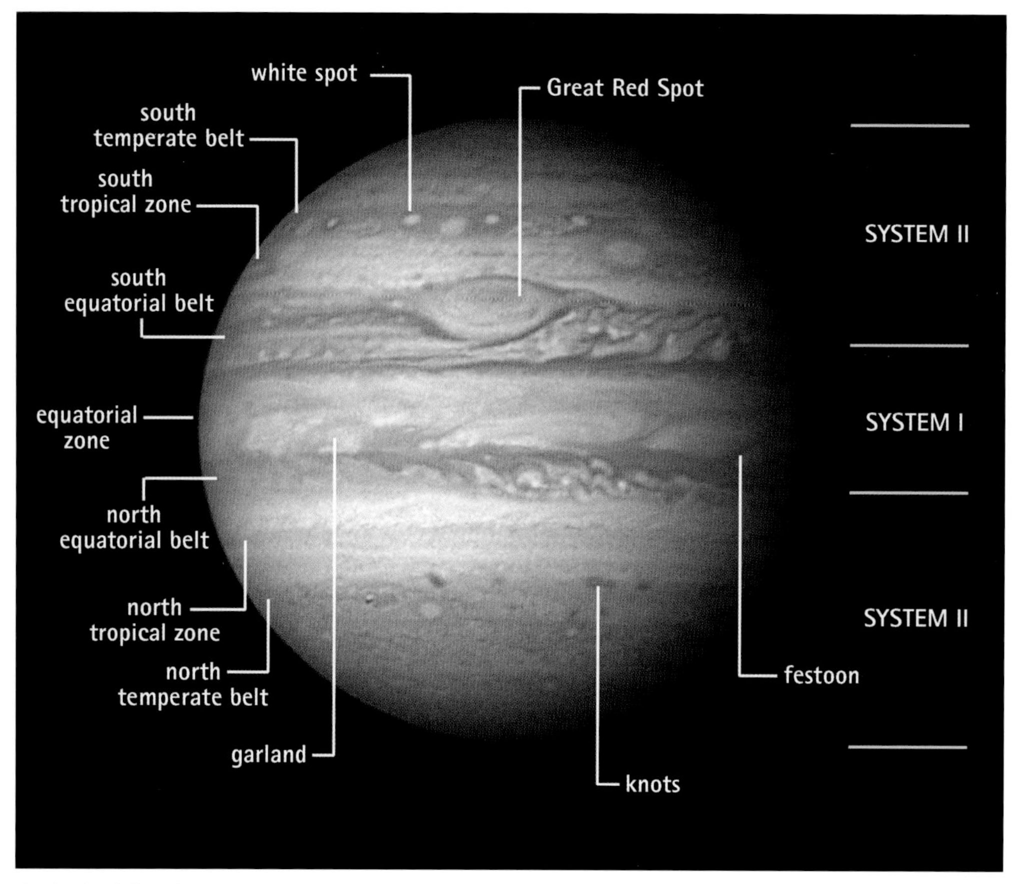


Jupiter's visible features are divided into areas called belts and zones. South is up.

equatorial bulging in a planet is an indicator of how fast it is rotating and since Jupiter, like our Sun, is a fluid-like, gaseous world the effect is more pronounced. Of all the planets, Jupiter's fast period of rotation is noticeable within minutes if one carefully monitors the progress of features across its disc. If you take a coffee break and return to the telescope 15 or 20 minutes later you will notice a dramatic change. Its dynamic internal activity and rapid spin results in a plethora of ever-changing cloud features that circle the planet at slightly different rates. The zone about the equator is referred to as System 1 and has the fastest daily passage of 9 hours, 50 minutes. Latitudes north and south of System 1 are referred to as System 2 regions. These regions have a typical diurnal period of 9 hours, 55 minutes. There is a third system based on the rotation of Jupiter's core but it is of no significance to visual observers.

## Features of the Jovian clouds

When we observe Jupiter all that we see are the topmost dense cloud layers. A small telescope easily reveals the greyish-looking cloud belts and whiter zones that encircle the planet. Larger aperture telescopes allow our eyes to better respond to the colour characteristics of these otherwise grey-looking features. The equatorial belts take on a subtle brownish hue and the regions between them (the zones) appear more yellowish-white. With apertures of 200mm (8in) and greater it is possible to discern, within the planet's equatorial zones, discrete wispy features called *festoons* and *garlands*. While not limited to this region, they are most commonly found here. Festoons appear to curve across the lighter coloured zones from one belt to another while garlands appear as small wispy loops or thin vertical protrusions extending from a belt into an adjacent zone. They are a regularly seen feature of Jupiter and may simply appear and disappear within days or weeks. Another feature known as a *column* appears like a vertical stanchion that extends across a zone joining one belt to another. Visually these three features are greyish or blue in colour, a robin's egg. They are testament to the extreme rifting activity along the borders of the belts and zones.

Frequently seen in the northern equatorial belt are small dark patches known simply as knots. These dense or darkened areas can at times look deceptively like a transiting moon. In contrast to knots, breaks or bright regions similar in colour to a zone sometimes appear along the rusty brown belts looking somewhat like a broken link. These are referred to as *gaps*.

Like the edge of a wood saw, *notches* are yet another feature that become visible from time to time. With a rippling appearance, they border the edge of an equatorial belt and the adjacent zone.

## A huge storm

The most well-known and active feature of Jupiter is a churning, oval-shaped anti-cyclone storm called the Great Red Spot (GRS). It has been present in Jupiter's atmosphere since it was first observed over 300 years ago but may not be a permanent feature. Situated in the southern System 2 region between the south equatorial and south temperate belts, it rotates within an area known as the Red Spot Hollow. If you're wondering what an anti-cyclone is, it is defined as an atmospheric system where the barometric pressure is high and the circulating air

These three tri-colour video camera images of Jupiter reveal just how much eruptive activity occurs in the planet's cloud tops. This sequence shows the planet's rotation within a couple of hours' observing. At left is the Great Red Spot. At centre, a large festoon arches across the equatorial zone while to its right is a garland. At right, several dark knots can be seen in the north equatorial belt while sawtooth-like notches appear along the southern edge of the south equatorial belt. South is up.

tends to move outward from its centre. An anti-cyclone rotates clockwise in the Northern Hemisphere and anti-clockwise in the Southern Hemisphere.

The overall size of the GRS shrinks and grows slightly from year to year. It has in the past been known to disappear almost completely for a short period. At its maximum, this titanic storm could engulf up to three Earths; it's hard to comprehend a storm of this size! Throughout most apparitions, however, the GRS can be seen but is not always red. In the last decade it has appeared much paler with a pink, orange or yellowish hue. The GRS and surrounding region is a place of great disturbance where turbulent cloud activity is nearly always seen leading and trailing it.

From time to time in this very active region, smaller storm systems called white spots develop from turbulent eruptions deep within the Jovian atmosphere. Circular or sometimes oval in shape, their movements and changes can often be monitored for long periods. Some white spots occasionally merge with other local cells and increase in size. Others can simply move dependently within a belt but may occasionally be drawn in and swallowed up by the intensely disruptive activity around the Great Red Spot. White spot activity is perhaps greatest in the mid to high southern regions of Jupiter, but it can also occasionally be found at similar latitudes in the Jovian Northern Hemisphere. You may at times be fortunate enough to witness the rifting disturbance of a white spot eruption within a belt, as it spreads out in longitude, over several days. Following the progress of such occurrences can be quite interesting, especially if the observer is equipped to photograph or video various stages of the event. Such opportunities make for a fine time lapse movie.

While observing these seemingly delicate loops and twists across Jupiter, take a moment to consider that some of these clouds are actually travelling at up to 500 kilometres per hour. A sobering thought indeed!

Jupiter's incredibly fast rotation offers a unique opportunity to record an entire rotation of the planet during the maximum dark hours of winter months, when oppositions occur in that season. Using a Wratten No. 21 (orange) or 12 (yellow) glass filter will greatly enhance the contrast of festoons, while a light blue filter such as a Wratten No. 80A will greatly improve belt contrast. A red Wratten No. 23 or 25 improves contrast around the perimeters of the Red Spot Hollow as well as festoon and garland activity.

Jupiter taken with a video camera and 200mm (8in) reflector telescope in 2003. The ever-changing Great Red Spot can be seen in the southern equatorial belt, appearing a more brown-orange hue in this image. Several smaller circular storms, called white spots, can also be seen in the southern temperate belt region above it. The two moons seen from left to right are Io and Europa.

# Jovian moons

While the ever-changing clouds twist and contort high in Jupiter's atmosphere, the four largest moons appear to flank the planet like a string of pearls. Each moon presents an apparent angular disc diameter of less than 2 arc seconds and with individual magnitudes ranging from 4.2 to 5.2 around opposition, they are easily imaged with a CCD or film camera. In fact, they could be seen easily with the naked eye were it not for Jupiter's dazzling brilliance. When the sky is incredibly steady and the seeing is superb, it is possible with a 150mm (6in) reflector to see the respective discs of each moon at higher powers, but this will most certainly be more comfortable with apertures of 200mm (8in) and greater. During opposition, their respective angular sizes range from between 1.1 arc seconds for Europa to 1.85 arc seconds for Ganymede.

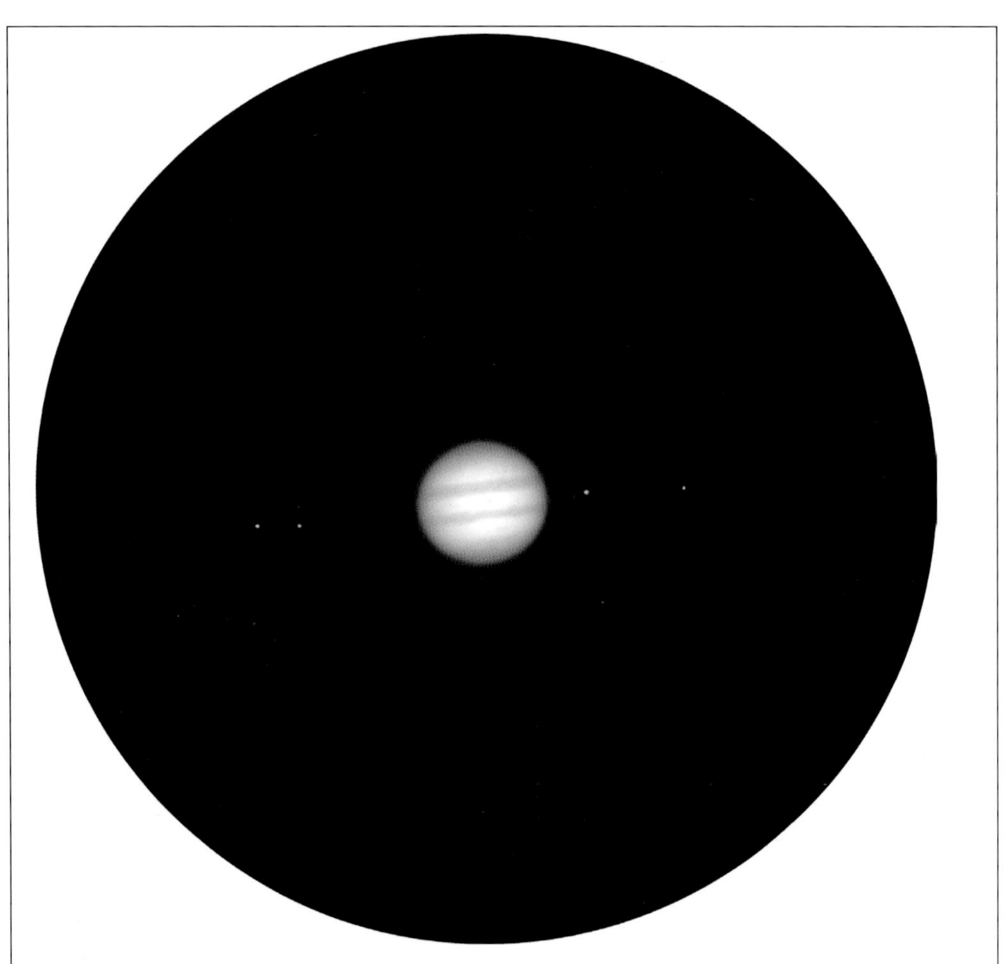

A typical low-power view of Jupiter and the four Galilean moons as can be seen at the eyepiece of most small- to medium-sized telescopes. Even in binoculars or a small beginner's telescope, these small and unique worlds can be seen to change positions within an hour. The brownish hue of the encircling belts is more easily noted in telescopes with apertures 100mm (3.9in) or larger.

This enlarged view of Ganymede shows how the moon's disc and subtle albedo markings at the surface can be captured on video with a 250mm (10in) telescope. The horizontal lines are video scan lines.

Visually distinguishing the colours of each moon can be a bit difficult and is, in my opinion, subjective. Although Jupiter's second-largest moon, Callisto, does exhibit a brownish-red hue, Ganymede and Io appear a more tan-orange colour. When compiling a tri-colour image of Jupiter however, the colours of its moons are more accentuated than can be detected visually.

Undoubtedly, the most appealing and observationally contrast-rich target is Ganymede. Since it presents the largest angular size of the four moons it is possible with a 250mm (10in) reflector to detect subtle surface markings. Ganymede's dark regions are greatly accentuated in contrast to surrounding brighter grooved terrain. It is also interesting to note that the bright lunar highlands of our own Moon are somewhat darker than the dark regions of Ganymede.

Observing or imaging fleeting glimpses of possible surface markings is difficult at the best of times. The best tool for capturing these random moments of steady sky is a video camera.

## Satellite transits

Sweeping back and forth in their orbits around the planet's equator, the Galilean moons occasionally cross the face of mighty Jupiter in magnificent transit events. Since the largest moons orbit along a relatively level plane about Jupiter's equator, and Jupiter itself orbits in roughly in the same plane as the Earth, we are well situated to observe these striking events. It truly is a marvellous encounter to behold when we can witness the circular shadow of a moon belonging to another world projected on its cloud tops. When a moon itself passes in front of a planet it is called a *satellite transit* and when its shadow is seen this is called a *shadow transit*.

Observing the shadow of a moon cast onto Jupiter's cloud tops is far easier than observing the moon passing across the disc of the planet itself, which can easily be lost in the background glare. Eyepiece filters can help discriminate low albedo transiting moons from the highly reflective Jovian atmosphere. Transit opportunities suggest a time lapse movie, whereby the ingress and egress motions of the satellite and its shadow are captured in several photos or camera images taken at staggered intervals.

Io, innermost of the four moons, maintains an orbital period of 1.769 days and, considering its close proximity to Jupiter, offers the most commonly recurring transits. With an orbital period slightly less than 4 days, icy Europa is the next most frequently transiting moon. The most striking of all are the shadows projected by the two more distant and larger moons, Ganymede and Callisto. Ganymede is in fact the largest moon in the Solar System, bigger even than the planet Mercury. Ganymede has an orbital period of 7.1 days while one Callisto orbit takes 16.7 days.

A much rarer and more exciting opportunity is a dual transit event where two moons and their respective shadows pass across the face of Jupiter at around the same time. Definitely not a time for your camera to fail!

# Occultations and eclipses

When a moon passes behind the disc of a planet it is occulted by the planet and when darkened by its shadow it is said to be eclipsed. Both of these events are amazing to observe. Because Jupiter is not a solid body like our Moon, the light from a star or one of its moons is not abruptly cut off from view once reaching the perceived visible edge of its limb. A moon appears to merge with or fade behind the semi-opaque limb of the upper atmosphere for a short period after initial contact. In fact, Jupiter exhibits a degree of limb darkening similar to the Sun because of its gaseous make up. A bright moon like Europa passing behind the Jovian limb often emphasises the difference in brightness.

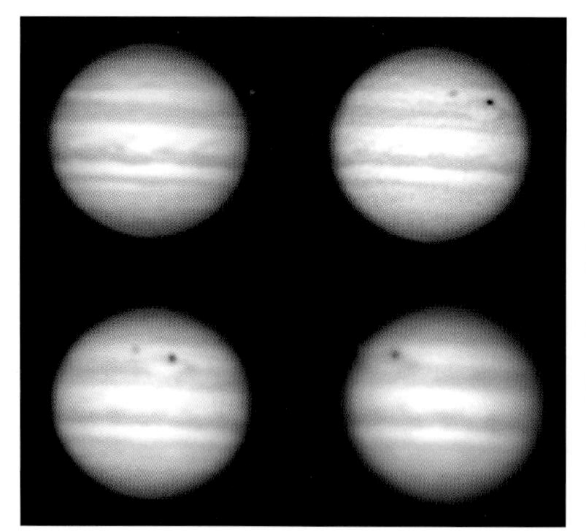

This sequence shows the transit of Ganymede in September 1998. Ganymede itself can be seen moving along the south tropical zone while the much darker shadow of the moon is projected onto the cloud tops of Jupiter's Great Red Spot. South is up.

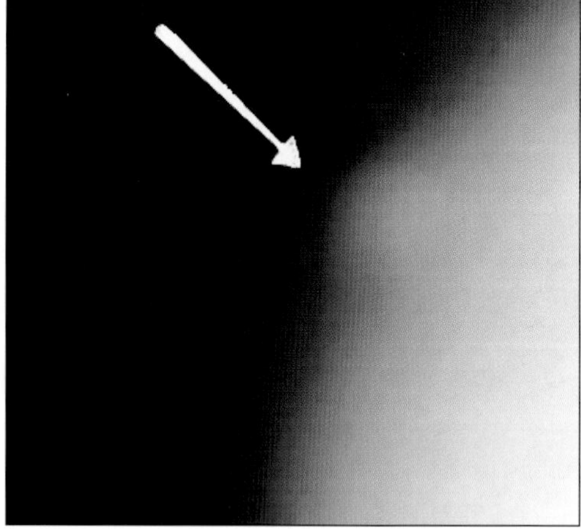

Ganymede disappearing behind Jupiter during an occultation in 1998. Since Jupiter is not a solid object like our Moon, Ganymede appears to fade away into the planet's gaseous semi-tranparent limb.

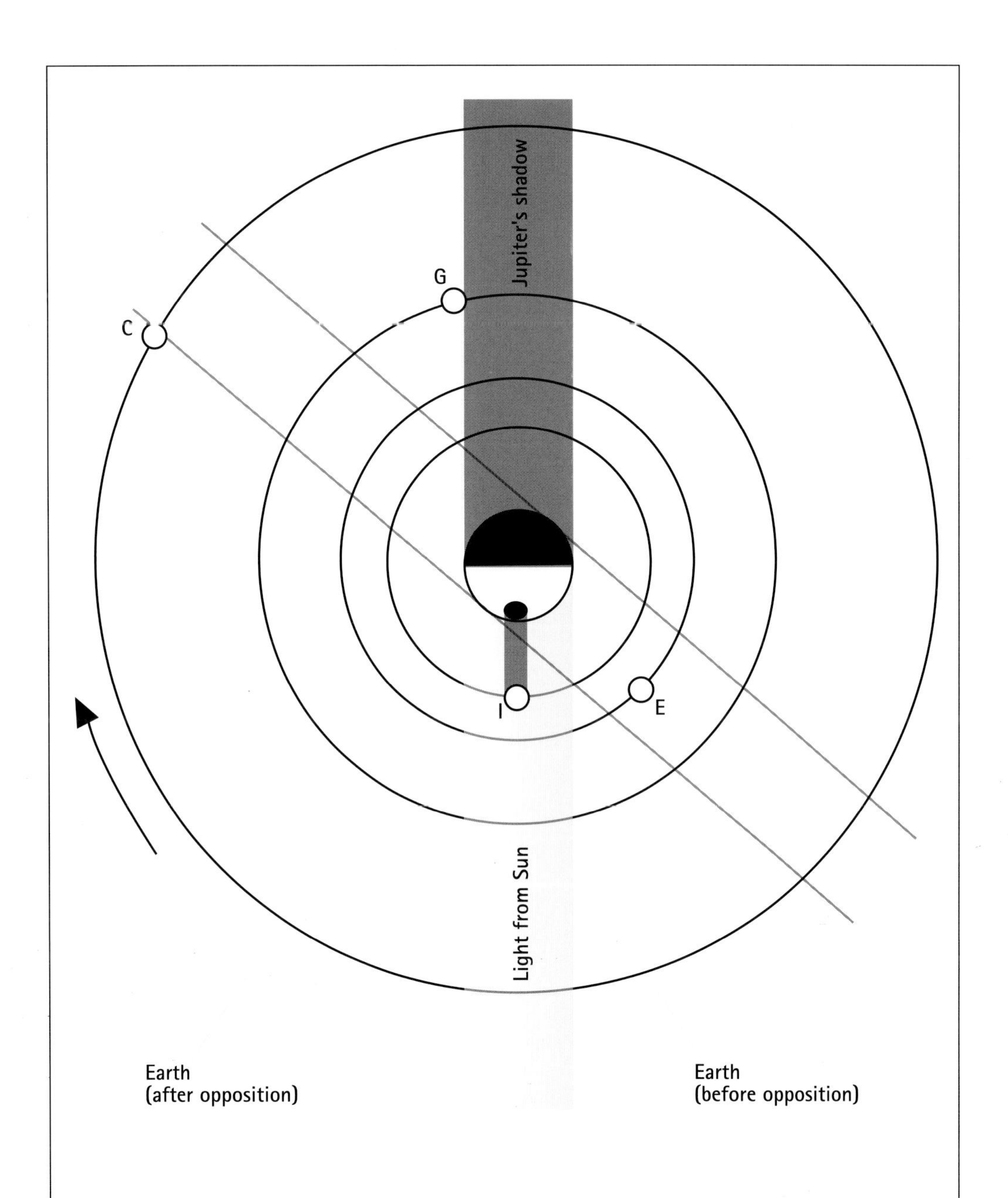


Aspects of transits and occultations of the Jovian moons. In the line of sight from Earth (before opposition example), Callisto (C) is being occulted by Jupiter while Ganymede (G) is about to be eclipsed by Jupiter's shadow. At the same time, the final stages in a shadow transit of Io (I) can be seen while the tiny disc of Europa (E) is still in mid transit across the face of Jupiter.

In the months leading up to or after opposition, when Jupiter's shadow is cast to the east or west, it is just as interesting to observe the predicted disappearance or reappearance of a moon as it moves into or emerges from darkness at some distance from the planet's limb. Europa's highly reflective, icy surface makes it the easiest to track whilst passing into Jupiter's shadow. To make observing the moon easier, use the highest practical magnification and move the telescope so that Jupiter's disc is partially outside the field of view. This will reduce obtrusive glare, allowing your eye to better adjust to the dimmed light of the moon being eclipsed.

# Daylight observations

Like Mercury and Venus, Jupiter can be observed during daylight hours. In fact the equatorial belts can be seen quite clearly and contrast can be further improved using a coloured filter such as a blue Wratten No. 80A or red 23A. A shadow transit of a moon across the face of Jupiter is also easily noted, though you will not be able to see the moon itself.

**Future dates of opposition**

4 March 2004
3 April 2005
4 May 2006
6 June 2007
9 July 2008
15 August 2009
21 September 2010

# Saturn

| | Saturn | Earth |
|---|---|---|
| Apparent visual diameter (arc seconds) | 18.4–20.7 | - |
| Axial tilt (degrees) | 25.33 | 23.45 |
| Equatorial diameter (km) | 120 536 | 12 756 |
| Surface gravity (metres per second$^2$) | 9 | 9.78 |
| Magnitude (brightness) at opposition | max. –0.3 min. +0.9 | - |
| Mean distance from Sun (km) | 1 429 400 000 | 149 597 870 |
| Maximum distance from Earth (km) | 1 658 854 980 | - |
| Minimum distance from Earth (km) | 1 195 772 020 | - |
| Number of known moons | 31 | 1 |
| Orbital inclination (degrees) | 2.488 | 0 |
| Orbital period (years) | 29.4 | 1 |
| Primary atmospheric composition | hydrogen | nitrogen/oxygen |
| Rotational period (d/h/m) | 0d10h14m | 0d23h56m |

Far beyond the orbit of mighty Jupiter, temperatures plummet to an icy –190°C where we encounter another formidable gas giant, the intriguingly beautiful ringed world of Saturn. The sixth planet from the Sun and second largest in the Solar System, Saturn also holds second place in terms of its mass, which is about ninety-five times that of Earth.

It orbits at nearly twice the distance of Jupiter from the Sun with a mean distance of 1 429 400 000 kilometres. Imagine you could take a journey from Earth to Saturn in the family car. When at its closest, and if travelling in a straight line at a non-stop speed of 100 kilometres per hour, it would take you around 1365 years to reach it. But modern spacecraft can make this journey in a matter of several years

A true-colour picture of Saturn taken from *Voyager 2*. Many newcomers are often inspired to buy a telescope when presented with spacecraft views like this. While Saturn is without doubt a wonderful telescopic object, most of the fine banding seen here on the planet's disc and within the rings is not visible through common small backyard telescopes.

by increasing their acceleration with a gravity-assisted fly-by of two or more inner planets, commonly referred to as a planetary slingshot. When we look at Saturn in the sky, the reflected sunlight light we see off its cloud tops takes over an hour to reach the Earth.

In 1655 Dutch astronomer Christiaan Huygens first noted the presence of Saturn's incredible rings. Within a few short years astronomers soon realised that this remarkable feature actually consists of several concentric rings. The first hint leading to this realisation came in 1675 when Italian astronomer Giovanni Cassini discovered a thin gap in the rings, now aptly named the Cassini Division.

With even the largest ground-based telescopes, Saturn's rings appear smooth, with only a handful of banded divisions. During the historic 1980 and 1981 encounters, the *Voyager* spacecraft revealed the long-suspected true nature of Saturn's ring system—a complex series of many finer island rings made up of millions of small icy rock particles and separated by small moonlets called shepherding moons. The overall extent of the rings from one edge to the other spans over 960 000 kilometres, encircling the planet's disc at the equator. The planet itself is comprised primarily of hydrogen (93 per cent) and helium (5 per cent) and, like Jupiter, is also thought to possess a highly compressed liquid metallic core. Similar in make up to Jupiter, Saturn also possesses belts and zones; these, however, are much less vivid due to a layer of ammonia-ice crystal haze suspended above the yellowish cloud deck. The winds below travel at very high speeds of up to 400 metres per second.

At the time of writing, Saturn has at least thirty-one moons that orbit the planet, yet only a handful are visible through small telescopes from Earth. The *Voyager* missions sent back stunning images and scientific information about the planet, its rings and several moons that will continue to keep planetary scientists busy for years to come.

The arrival of the current *Cassini* mission to the planet is highly anticipated and, like the *Galileo* spacecraft at Jupiter, it will be placed in an orbit around Saturn, allowing protracted studies for many years. The *Cassini* spacecraft carries with it a small probe called *Huygens*, which, it is planned, will descend onto the surface of Saturn's largest moon, Titan.

Titan is the second largest moon in the Solar System, slightly larger than the planet Mercury. Like cloud-covered Venus, Titan's surface has eluded our views, due to the thick hazy atmosphere which enshrouds this mysterious satellite world. If the *Huygens* probe survives its descent and landing, either on a liquid or solid surface, we can expect some very exciting results.

## Observing Saturn

Saturn is perhaps the most well known of the planets because of its striking ringed appearance. There is little doubt that, for aesthetic appeal, Saturn is the jewel of the sky. The first time one witnesses this almost surreal world in a telescope is something not easily forgotten, leaving even the most conservative individuals expressing a breath-like 'wow!'.

Speaking from personal experience, Saturn is perhaps the one planet that stirs the most visual and emotional senses, particularly when observed in increasingly larger telescopes. Through a 0.6m (24in) telescope in steady, pristine skies, the visual magic of this ringed island world suspended in the vastness of space is an overwhelming experience that a static picture cannot replicate. While the planet itself is physically smaller than the globe of mighty Jupiter, it is still an incredibly giant world. Saturn's disc spans roughly 9.5 Earth diameters, or 120 536 kilometres at its equator. Even with its second-place status in the giant stakes, Saturn's apparent visual size as seen through a telescope is further diminished since it orbits nearly twice as far from Earth as Jupiter.

When you first observe Saturn, especially through a small telescope, don't expect it to appear like the commonly seen glorious pictures in books or misleading decorative packaging of economical beginner's telescopes. While it is an awesome planet to behold, at the eyepiece its yellowish disc spans little more than 20 arc seconds around times of opposition, gaining something in overall

The main features of Saturn visible through small- to medium-sized telescopes are labelled in this diagram.

size with the rings spanning 45 arc seconds. A small telescope will reveal the planet and its rings easily at around 40X power but larger apertures with greater resolving powers present much more exciting views. If the sky is steady, a 200mm (8in) telescope or larger will produce magnificent views at 200X to 300X power. If sky conditions are near perfect, try bumping up the magnification further without the image becoming unreasonably dim. But be warned—you may become addicted!

Like Jupiter, Saturn exhibits noticeable flattening of its disc at the poles due to its fluid-like atmosphere and rapid rotation. When the planet and its rings are observed edge-on, this flattening is quite obvious. Saturn's rotational period is not uniform

This colour image of Saturn was taken from suburban Sydney in early 2003, using a 200mm (8in) Newtonian reflector operating at f/30. It is a composite of three images taken using a low-cost black-and-white video camera and individual RGB colour filters. Note the brightness differences between the outer A ring and inner B ring. The dark Cassini Division separates the two and the hazy inner C ring can also be seen.

across the entire globe because, like the Sun and Jupiter, it has no solid surface. Like the belts and zones within Jupiter's atmosphere, Saturn also has differential rotation at various latitudes. About the equator the period is quoted as 10 hours, 15 minutes while at the temperate regions, plus or minus 35°, the period is around 10 hours, 38 minutes. At higher latitudes the rate is slightly longer still.

A 150mm (6in) telescope reveals the dark polar hood, brownish bands and bright equatorial region with little difficulty. More often than not, definition between the belts and zones appears as little more than shades of yellow and brown. A Wratten No. 80A light blue filter will improve contrast while not oversaturating the image with a complete blue tone, thus the brownish colours are still apparent. Like big brother Jupiter, Saturn occasionally (though rarely) features transient white ovals and spots. Such phenomena have been observed by a number of amateurs over the years, and are known to last for several days or weeks, but lack the seemingly permanent nature of an enigma such as the Great Red Spot of Jupiter. These spots can be seen moving around the planet's disc as it rotates. Some patches or spots simply fade away, while others have been known to stretch out into long white streaks, eventually fading into the surrounding atmosphere.

If you are keen to undertake any serious observations of Saturn in the hope of detecting such phenomena, a telescope of 200mm (8in) aperture or larger will be best suited to the task. For most of the time, however, Saturn's disc appears rather bland, with the exception of the subtle gradations in banding mentioned earlier. With this in

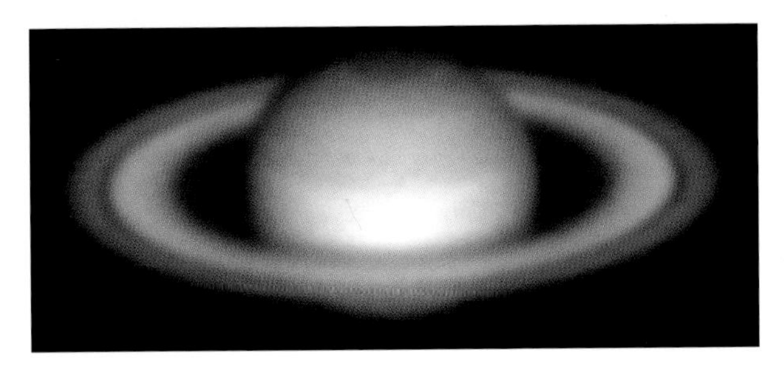

Although atmospheric phenomena such as white spots are rarely seen on Saturn, there are occasionally other developments such as this transient banded red collar just below the polar hood, photographed in 1999.

mind there is no great emphasis on timing when it comes to making a sketch or tri-colour picture of the planet. On the other hand, if you should detect an unusual marking then the timing of your drawing or image becomes somewhat more critical since the suspect feature will exhibit a change in position relatively quickly.

When drawing the planet, get the important information down first. Sketch the basic appearance of the disc then the noted feature. You can then complete the more aesthetic information later, such as shades of gradation from the polar hood to the equator or obvious belts and zones plus details in the rings. You should try to make several sketches or images of the planet noting the universal time for each, as this will be useful to others who may want to confirm your observations. Another unusual development to watch for is the occasional reddening or darkening of the cloud bands (like a collar) near the bluish, dark green polar hood, as occurred in 1999.

## The magical rings

The rings are certainly the most mesmerising feature of Saturn. Even through a small telescope the most striking visual treat is the almost 3D effect of the planet's shadow cast across the rings when at eastern or western quadratures. Each year we can observe the rings from a different perspective because the planet is titled by almost 27° to the plane of its orbit. During its 29.4-year passage around the Sun we occasionally look down on its rings while at other times we see them from the underside.

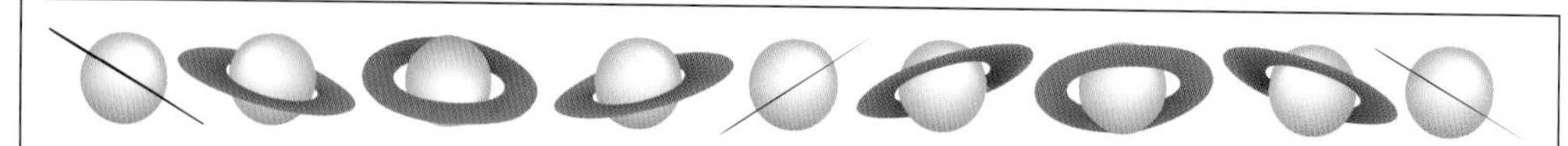

Due to Saturn's axial tilt to the plane of its orbit, we have the opportunity to observe the poles and rings from different perspectives: above, edge-on and below.

Every 15 years, the rings appear edge-on as seen from Earth. Through a telescope they can appear almost invisible for a short period. This is not surprising considering the small, loose rocky bodies that make them up are confined to a plane less than 1 kilometre thick and possibly even as thin as 200 metres.

The Cassini Division is the thin dark gap that separates the prominent outer 'A' and inner 'B' rings. With larger instruments and excellent seeing conditions, it is possible to detect the very delicate and elusive Encke Division near the outer edge of the A ring. This can present a visual challenge if your telescope has poorly collimated optics so be sure to check alignment before observing.

Using a red Wratten colour filter number No. 23A or 25 will improve contrast between ring divisions. With careful scrutiny, you should be able to detect the innermost ring, known as the Crepe or 'C' ring. Its gossamer appearance is very faint compared to the bright neighbouring B ring and is seen as a hazy arc where it crosses the disc of Saturn. With a keen eye through apertures 200mm (8in) or larger it is possible to detect colour differences between the three rings as opposed to differences in brightness. The innermost C ring has a slight red-brown hue while the outermost A ring exhibits a slight bluish-brown tinge. The B ring appears more pure white by comparison.

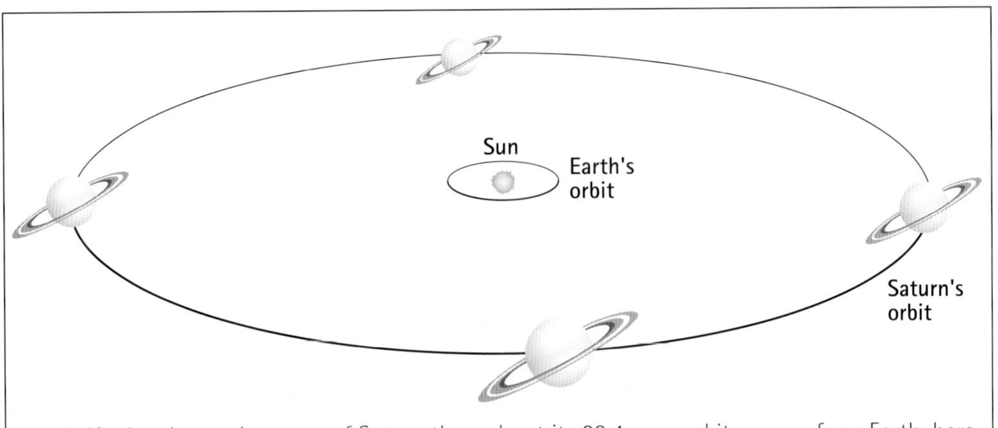


To clarify the observed aspects of Saturn throughout its 29.4-year orbit as seen from Earth, here the planet's 25° axial tilt can be seen relative to its orbital plane, which is very close to that of the Earth.

## Saturn's moons

Accompanying the stand-alone beauty of this planet are the numerous faint and tiny moons surrounding it. At magnitude 8.1 its largest moon, Titan, is easily visible through a small beginner's telescope. There are five other moons that can be glimpsed with larger aperture instruments. Rhea and Iapetus are the next brightest, however Iapetus does vary in brightness at certain times along its orbit due to darker surface material in one hemisphere compared with the other.

With its albedo varying from 0.05 to 0.5, the resulting effect is seen as alternating levels of brightness.

Tethys is the next brightest satellite, followed by Dione at around magnitude 10.6. Much fainter Enceladus has an apparent visual magnitude of around 11.8 but orbits fairly close to Saturn compared to the others, making it more difficult to spot. With good seeing conditions in dark skies you should be able to detect it with a 150mm (6in) reflector. At times it may only be possible to see the fainter moons using *averted vision*. Gasp! What is averted vision I hear you ask? Averted vision is the intentional use of our more sensitive peripheral vision, the visual perception of things surrounding that which we are looking at directly. The retina of the human eye is made up of cones at its centre which primarily respond to colour. The surrounding area is made up of receptors called rods and these are some forty times more sensitive than cones, but lack the ability to decipher colours. The intended use of this low-light, sensitive peripheral vision is well known among deep-sky observers for detecting very faint galaxies or nebulae otherwise invisible with direct vision. By simply looking slightly away from a subject, its presence becomes more obvious. Sometimes an object can seem to jump out at you from nowhere. This same technique can be applied to detecting the fainter moons of Saturn especially when using smaller telescopes. We may see a moon while focused on the planet itself but when turning our attention to the moon directly, it seems to disappear from view. A 200mm (8in) instrument or larger provides greater ease of visual detection of these five fainter moons.

## Plotting the moons

When at the telescope you can plot the positions for each suspected satellite on a piece of paper, noting directions of north and east. While plotting each star and moon in the field of view, be careful to draw their approximate scales of distances and angles relative to Saturn as accurately as possible. This will help later when referring to your computer planetarium software for positive identification. Note directions on your drawing to help with comparative orientation to the simulated object displayed on the computer. Also, place a number next to each suspected moon as an approximate ranking of brightness compared with Titan which, being the brightest, will be labelled 1. Where there are bright or known stars in the field of view you can indicate these by drawing a star-like symbol such as an asterisk (*).

The innermost visible moons of Saturn present noticeable change in their respective positions within a matter of hours. But to note an obvious change in Titan's orbit it is best to observe it over consecutive evenings.

You might like to consider making several drawings of Saturn and its moons over consecutive evenings with the view to constructing a time lapse animation after scanning these drawing on to a computer. To do so, you should use the same eyepiece/telescope combination each night in order to maintain the same field of view and magnification represented in each drawing. The orientation of Saturn should also be kept as close as possible to the drawing done each previous night, as this will help when aligning scanned pictures later. Of course, this is not overly critical since scanned images can be digitally re-oriented using an image-processing tool like Adobe Photoshop. The most important thing is to ensure that you maintain accurate positional distance scales for each of the moons, as this will be the most visible effect when the animation is running. All the images for the movie must be of the same pixel dimensions (e.g. 320 x 240 pixels). Once the sequence has been compiled you can then create the movie using a multitude of movie/animation programs available in formats such as mpeg, avi or gif. Once you have completed your masterpiece, you can display your movie as an animated gif file on your personal website to show off to the world.

This overexposed image of Saturn and its rings was required to reveal five of the planet's moons. Clockwise starting from the top is Titan, Tethys, Enceladus, Rhea and Dione. The image was taken from suburban Sydney using a 200mm (8in) reflecting telescope.

| **Future dates of opposition** |
| --- |
| 13 January 2005 |
| 28 January 2006 |
| 10 February 2007 |
| 24 February 2008 |
| 8 March 2009 |
| 22 March 2010 |

# Uranus

| | Uranus | Earth |
|---|---|---|
| Apparent visual diameter (arc seconds) | 3.6–3.9 | - |
| Axial tilt (degrees) | 97.85 | 23.45 |
| Equatorial diameter (km) | 51 118 | 12 756 |
| Surface gravity (metres per second$^2$) | 7.8 | 9.78 |
| Magnitude at opposition | max. +5.5 min. +6.1 | - |
| Mean distance from Sun (km) | 2 870 990 000 | 149 597 870 |
| Maximum distance from Earth (km) | 3 159 769 980 | - |
| Minimum distance from Earth (km) | 2 582 694 020 | - |
| Number of known moons | 21 | 1 |
| Orbital inclination (degrees) | 0.774 | 0 |
| Orbital period (years) | 84 | 1 |
| Primary atmospheric composition | hydrogen/helium | nitrogen/oxygen |
| Rotational period (d/h/m) | 0d17h14m | 0d23h56m |

Orbiting at mind-boggling distances from our Sun we encounter the ice worlds of our Solar System, where the Sun appears as a small disc or an extraordinarily bright star. Uranus was discovered on 13 March 1781 by Sir William Herschel. This discovery virtually doubled the extent of the known Solar System at that time. Unlike the other planets, Uranus is tipped on its side, perhaps caused by some catastrophic event early in its history. It was also discovered to possess a ring system of its own in 1977 when astronomers watched the faint light from a distant star wink on and off as each ring passed in front of it.

Uranus is a gas giant like Jupiter and Saturn although it is often referred to as an ice world. When *Voyager 2* encountered the planet in 1986 it revealed a disappointingly bland view. The Uranian atmosphere is comprised mainly of hydrogen and a smaller percentage of helium while its greenish-blue colouring is attributed to an upper atmosphere of

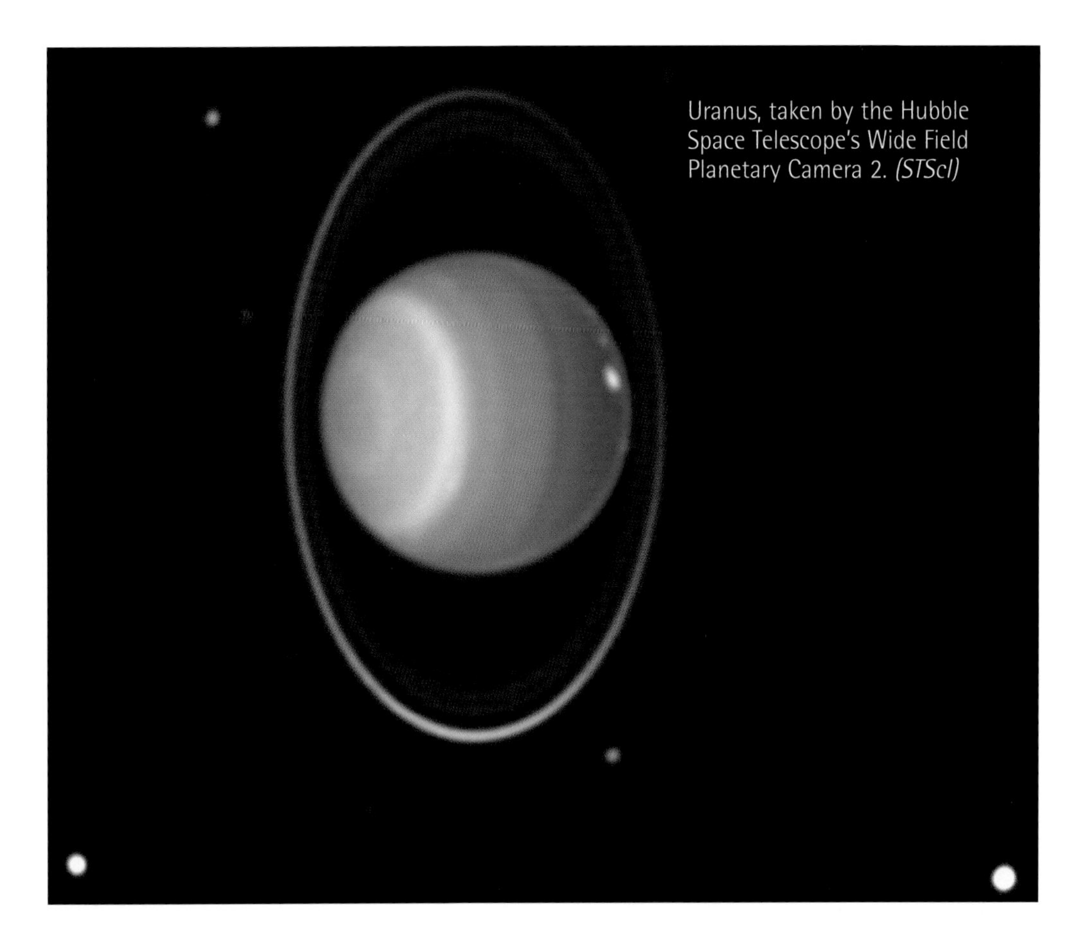
Uranus, taken by the Hubble Space Telescope's Wide Field Planetary Camera 2. *(STScI)*

methane. Although stunning images of its ring system were returned by *Voyager 2*, the planet itself revealed little detail. But like some divine compensation, the moons of Uranus proved to be geological gold mines. Most peculiar is Miranda, with perhaps one of the most diverse surface terrains in the Solar System. Peppered with old cratering it features streaked and irregularly contorted regions that perhaps indicate it was ripped apart in its early history then later reassembled itself through the forces of gravity. Its indelible scarring is possibly testament to the catastrophic event that tipped Uranus on its side.

Uranus has twenty-one natural satellites, some of which were discovered by *Voyager 2* in 1986. The four largest are Titania, Oberon, Umbriel and Ariel respectively. Each of these moons has a diameter exceeding 1000 kilometres. Miranda, the fifth largest, was discovered in 1948 by Gerard Kuiper and is a sizeable step down in equatorial diameter, compared to the other four moons, at only 480 kilometres.

# Observing Uranus

Uranus takes 84 Earth years, or an entire human life span, to make one orbit of the Sun. By contrast, its day lasts only 17.2 hours.

On a clear moonless night Uranus can be spotted with the naked eye, although you will need to know exactly where to look; with binoculars, a modest star chart and a current planetary ephemeris it can be located without too much difficulty. Some ephemeris publications and computer software offer useful benchmark times for locating the planets, indicating when they may be closely situated in the sky with more easily identifiable targets like the Moon or a bright star. Due to its great distance from the Sun, Uranus appears to move slowly among the stars, trekking roughly a quarter of a degree over 10 days.

A telescope of around 100mm (3.9in) aperture or greater will distinguish its tiny greenish-blue disc at about 150X magnification, revealing its planet-like appearance compared with nearby stars. Even at opposition, Uranus is greater than 2.5 billion kilometres from Earth, sporting an apparent disc size of less than 4 arc seconds. This is only slightly greater than the apparent size of Mars when farthest from Earth. Since Uranus is so distant and physically smaller than Jupiter and Saturn, defining its disc sharply depends on how good the atmospheric seeing is at the time of observing. Since *Voyager*'s cameras confirmed its featureless nature, an observer at the eyepiece can expect to achieve little more than confirmation of its disc. At rare times of superbly clear and stable skies it is always a challenge to increase telescopic powers in order to view the planet's disc as large

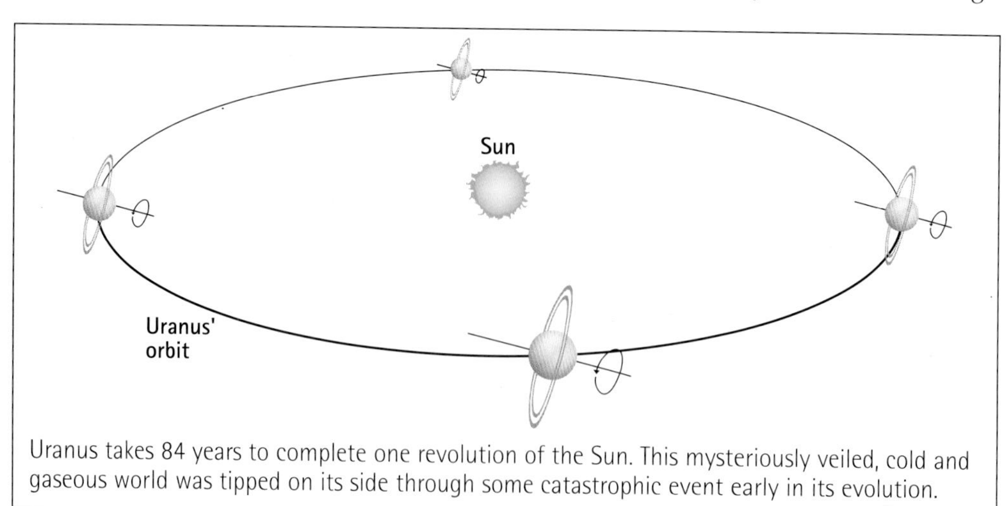


Uranus takes 84 years to complete one revolution of the Sun. This mysteriously veiled, cold and gaseous world was tipped on its side through some catastrophic event early in its evolution.

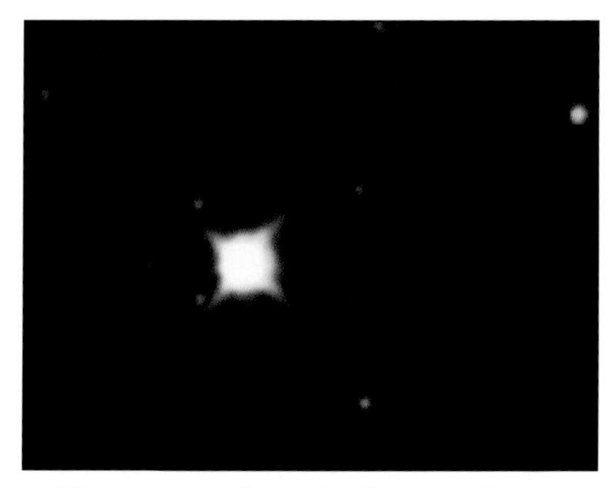

Uranus taken in 1985 with a 250mm (10in) reflecting telescope shows two 14th magnitude moons that orbit the planet in this 2-minute exposure taken with ISO 400 film. Just above the top left diffraction spike is Oberon, while Titania is the point of light near the bottom-left spike. *(Courtesy Steve Quirk)*

as is practicable before the image starts to noticeably degrade. Simply logging Uranus in your records as another planet observed may be enough to satisfy.

There are other challenges, however, should you wish to follow the path of Uranus across the night sky throughout the year. Of its several moons, only three are within the light-gathering power of most moderately-sized telescopes. One of the oddities to remember when observing the Uranian moons is that the planet's poles and its system of satellites are tilted almost to the plane of its orbit. Thus, the moons are not seen orbiting edge-on to the plane of the ecliptic as we see with the moons of Jupiter. When the poles of Uranus are pointed towards the Earth, its moons orbit about it like the outlying numbers on the circular face of a clock with Uranus at the centre. Further along its orbit the planet's equator eventually faces earthward and its moons appear to move up and down in a vertical motion.

Although Uranus is small and relatively dim at the eyepiece it can still present an obscuring glare when you are attempting to detect the nearby 14th magnitude moons. Positively identifying Titania, Oberon and Ariel requires careful scrutiny. A guiding eyepiece with cross hairs centred over the disc of Uranus can be useful in helping to reduce the glare of the planet. Tracked over the course of a few nights, the moons' positional changes can be recorded in a sketch, CCD image or carefully tracked photography. You can then identify which moon is which later, by referring to a current ephemeris or similar software tool on your computer.

**Future dates of opposition**

1 September 2005
5 September 2006
9 September 2007
13 September 2008
17 September 2009
21 September 2010

# Neptune

| | Neptune | Earth |
|---|---|---|
| Apparent visual diameter (arc seconds) | 2.49–2.52 | - |
| Axial tilt (degrees) | 28.3 | 23.45 |
| Equatorial diameter (km) | 49 527 | 12 756 |
| Surface gravity (metres per second$^2$) | 11 | 9.78 |
| Magnitude at opposition | max. +7.6 min. +7.7 | - |
| Mean distance from Sun (km) | 4 504 300 000 | 149 597 870 |
| Maximum distance from Earth (km) | 4 686 510 980 | - |
| Minimum distance from Earth (km) | 4 306 660 020 | - |
| Number of known moons | 11 | 1 |
| Orbital inclination (degrees) | 1.774 | 0 |
| Orbital period (years) | 165 | 1 |
| Primary atmospheric composition | hydrogen/helium | nitrogen/oxygen |
| Rotational period (d/h/m) | 0d16h6m | 0d23h56m |

Travelling ever further out into deep, cold space we come to Neptune, the eighth planet from the Sun. The discovery of Neptune was quite a feat of human observation and mathematical endeavour. After astronomers had noted irregular variations in the orbit of Uranus, two mathematicians, John C Adams and Urbain Le Verrier, calculated possible positions for the then undiscovered new world. Recent reports in 2003 indicated that Le Verrier is perhaps more deserving of a larger slice of credit for more accurately plotting positions subsequently leading to Neptune's discovery. And so it was in 1846 that JG Galle and Heinrich D'Arrest used these predictions to ultimately find Neptune. This is a world so far away that it takes almost 165 years to complete an orbit of the Sun.

In the last of the *Voyager* spacecraft missions, *Voyager 2* took 12 years to reach Neptune. The transmissions

Neptune, taken from *Voyager 2* at a range of 14.8 million kilometres on 14 August, 1989. *(NASA/JPL)*

from the spacecraft were so weak that it took every bit of radio dish receiving power here on Earth to discriminate its faint signal as distinct from all other background radio noise. This includes all those everyday signals such as television and radio which bombard us daily. Travelling at the speed of light, the transmissions took over 4 hours to reach the Earth.

After the visual disappointment of Uranus, Neptune was an overwhelmingly wonderful ending to an otherwise long mission. Although Neptune is seventeen times the mass of Earth, its blue coloured atmosphere and white streaks of cloud bears an uncanny resemblance to Earth—a sort of symbolic picture of home. A strong absorber of red light, the small amount of methane gas in Neptune's atmosphere gives the planet its pale blue colour. The *Voyager 2* images also revealed a great dark spot in its atmosphere similar to the Great Red Spot of Jupiter. Atmospheric wind speeds were recorded exceeding 1600 kilometres per hour.

At the time of writing, Neptune has eleven known moons of which Triton is the largest. Slightly larger than Pluto, Triton orbits Neptune at an average distance of 354 760 kilometres and completes one revolution in 5.87 days. Oddly enough it has a retrograde orbit which means it orbits the planet in the opposite direction to all the others. Triton is a cold world in the extreme and its thin nitrogen atmosphere freezes into a solid ice cap during winter. Pictures returned from *Voyager 2* during its fly-by revealed elongated dark plumes on Triton's surface. They are believed to be geysers or volcanic vents where warmer liquid nitrogen, about 30 metres below, is pushed up to the surface and released as a gas containing dark carbon-like particles that are windswept across its surface. Later investigations of the returned data revealed that Triton is still an active world despite previous theories to the contrary.

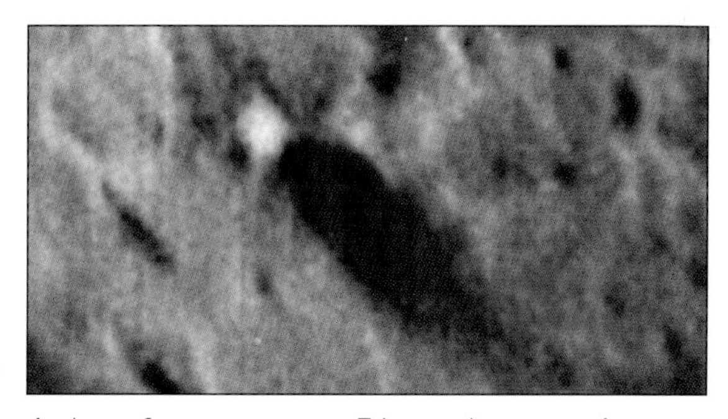

A plume from a geyser on Triton, taken on 25 August 1989 when *Voyager 2* was about 190 000 kilometres from Triton's surface. *(NASA/JPL)*

# Observing Neptune

Neptune is a faint target, too faint to be detected with the naked eye, reaching a maximum magnitude around 7.6 during opposition. Under a good dark sky, binoculars will reveal its faint bluish star-like appearance. Due to its great distance from the Sun, Neptune appears to move slowly among the stars, trekking roughly a quarter of a degree every 19 days. A good star chart and ephemeris is needed in order to positively identify it among the surrounding stars. Through a telescope at low power, Neptune appears a more stable point of light compared to other nearby stars with a bluish grey colour. Although it is a giant world in its own right, its extreme distance from Earth makes Neptune a difficult target to resolve in terms of a planetary disc. Around times of opposition Neptune appears no more than 2.5 arc seconds in diameter and requires a good quality telescope with an aperture of 150mm (6in) or greater to resolve it. To increase your chances of success, pick a night under very clear skies when the planet is near the meridian and the atmospheric seeing is very good.

Although Neptune is some 1.5 billion kilometres more distant than Uranus, Triton (slightly smaller than our own Moon) is about 13th magnitude in brightness, making it somewhat easier to detect than the much nearer largest moons of Uranus. Perhaps no less worthy than a photograph or CCD picture of Uranus and its moons, a carefully guided portrait of Neptune and Triton should certainly make you proud.

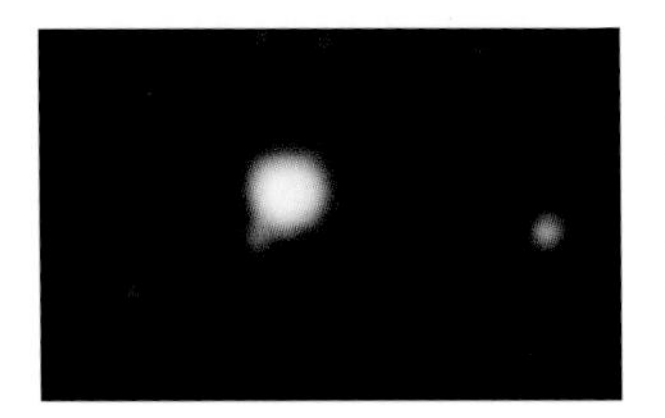

Even with a relatively small telescope in your own backyard you can view a planet and its moon over 4.5 billion kilometres away. This picture of Neptune and its largest moon, Triton, was taken in July 1985. Triton is the protrusion at bottom left edge of the overexposed disc of Neptune. *(Courtesy Steve Quirk)*

**Future dates of opposition**
8 August 2005
11 August 2006
13 August 2007
15 August 2008
18 August 2009
20 August 2010

# Pluto

| | Pluto | Earth |
|---|---|---|
| Apparent visual diameter (arc seconds) | 0.07–0.1 | - |
| Axial tilt (degrees) | 122.5 | 23.45 |
| Equatorial diameter (km) | 2 320 | 12 756 |
| Surface gravity (metres per second$^2$) | 0.4 | 9.78 |
| Magnitude (brightness) at opposition | max.+13.6 min.+16.0 | - |
| Mean distance from Sun (km) | 5 913 520 000 | 149 597 870 |
| Maximum distance from Earth (km) | 7 676 691 980 | - |
| Minimum distance from Earth (km) | 4 283 023 020 | - |
| Number of known moons | 1 | 1 |
| Orbital inclination (degrees) | 17.148 | 0 |
| Orbital period (years) | 248.5 | 1 |
| Primary atmospheric composition | methane | nitrogen/oxygen |
| Rotational period (d/h/m) | 6d9h17m | 0d23h56m |

After the hugely successful discovery of Neptune in 1846, logic suggested that perhaps another world may orbit even further out in the coldest and most forbidding regions of the Solar System. By the early 20th century more astronomers were focused on celestial subjects way beyond the realms of the heliosphere (the spherical realm within which the solar wind and magnetic field of the Sun have an influence). Concerned with better understanding the structure of our expanding universe, the huge distances that separate the island galaxies from our own and the compositions of the stars, less attention was aimed at the possibilities of finding more planets. But during these years of waning interest in planetary studies, a few dedicated observers continued their systematic

An extremely clear view of distant Pluto and its moon Charon, taken by the Hubble Space Telescope's Faint Object Camera. *(Courtesy R Albrecht, ESA/ESO and NASA)*

observations. As with the successful discovery of Neptune using mathematical calculations, some astronomers also applied this predictive methodology in locating the ninth planet, but failed.

At Lowell Observatory in Arizona, Clyde Tombaugh, a former assistant to Percival Lowell, maintained a diligent photographic survey of the night sky in the hope of detecting the elusive planet. Just like the method still employed today in locating comets, faint meteoroids and asteroids, Tombaugh employed the blink comparator method, involving the rapid switching back and forth between two pictures taken of the same area of sky in order to detect any small changes in the positions of the 'stars'. In 1930 his hard work paid off and he indeed found the ninth planet, which was named Pluto.

At the time of writing, no space probe has yet visited the planet Pluto. With even the largest telescopes in the world, Pluto's disc is difficult to resolve. However, sophisticated adaptive optics coupled with large state-of-the-art ground-based telescopes are now able to resolve the planet's tiny disc along with orbiting companion Charon, which is slightly smaller. Even smaller than our Moon, it's not hard to understand why Pluto was so difficult to find, considering it has an average distance from the Sun of 5.9 billion kilometres!

Even more so than Mercury, Pluto's orbit is highly eccentric—so much so that it occasionally passes slightly inside the orbit of Neptune. Another oddity is its highly inclined orbit of 17° to the plane of the ecliptic. These two peculiarities, along with Pluto's diminutive size, have lead some astronomers to debate its true status as a planet since it exhibits more comet- or asteroid-like characteristics. Adding further to the enigma that is Pluto, was the discovery of its only known moon, Charon, in 1978. The Earth itself stands out among the other moon-bearing worlds in that our Moon is so large in comparison to Earth. Charon, though, is more than half the size of Pluto, orbiting it at a distance of 19 400 kilometres in a little over 6 days, 9 hours.

These photographs of Pluto, seen as a mere point of light, were taken 8 days apart in 1985, and show its extremely slow passage across the background stars. Pluto is probably more neglected than Mercury, since it is a very difficult planet to observe and requires a large-aperture instrument. You'll most certainly need a detailed star chart to positively identify it. *(Courtesy Steve Quirk)*

## Observing Pluto

For a beginner, observing or even finding this planet is no easy task. Even during its closest approach with the Earth it achieves a visual brightness of little more than 13.8 magnitude and positive identification requires the use of detailed star charts and a telescope of at least 200mm (8in) aperture to glimpse its faint light. Telescopes of larger aperture make this task somewhat easier. As seen from Earth, Pluto's apparent angular disc size varies between 0.1 to 0.07 arc seconds.

Capturing Pluto with smaller instruments can still be achieved but requires a method of building up a picture over time to reveal that which the eye cannot see. Long-exposure film photography is one such method but is gradually being abandoned for more efficient tools like cooled CCD cameras. Unlike conventional film cameras, astronomical CCD cameras have highly sensitive pixels, which can more easily register the faint reflected light from distant Pluto in a fraction of the time it takes with grainy high-speed films.

As mentioned earlier, Clyde Tombaugh used a blink comparator method to discover Pluto. The process involves rapidly flicking between two photographs of the same star field, or area of sky, taken a day or so apart. Any vagabond object in the photographs will appear to shift in position with respect to the otherwise fixed background stars. This can be done today with digital images taken from an astronomical CCD camera using a computer and most commonly available image-processing programs. Of course you can still take conventional photographs and digitise them to a computer using a low-cost image scanner. Adobe Photoshop is one program that can be used to switch rapidly between two consecutive layered images. To ensure that the background stars in each picture appear fixed from one image to the next, both images must be oriented for correct registration with each other. This method is also useful in detecting faint asteroids and distant comets.

| Future dates of opposition |
| --- |
| 14 June 2005 |
| 17 June 2006 |
| 19 June 2007 |
| 21 June 2008 |
| 23 June 2009 |
| 25 June 2010 |

# The vagabond worlds

## Asteroids

| Asteroid | Diameter (kms) | Orbital period (years) | Mean distance from Sun | Rotation period (hrs) | Magnitude (max. brightness) |
|---|---|---|---|---|---|
| **Ceres** | 960 x 932 | 4.60 | 413.6 | 9.08 | 6.9 |
| **Pallas** | 570 x 525 | 4.61 | 414 | 7.81 | 6.3 |
| **Vesta** | 530 | 3.63 | 353 | 5.34 | 5.2 |
| **Juno** | 240 | 4.36 | 400 | 7.21 | 7 |

(Note: Maximum brightness of an asteroid does not necessarily coincide with opposition.)

Sometimes referred to as the minor planets, asteroids are the rubble fragments left over from the formation of our Solar System. With often highly elliptical orbits, most are found within a region between Mars and Jupiter called the asteroid belt. This belt is home to possibly hundreds of thousands of asteroids ranging from as large as 940 kilometres in diameter to under 1 kilometre across. Around 20 000 or so have been numbered and another 15 000 are waiting to be officially acknowledged.

Asteroids have been observed with telescopes since the early 1800s, and the name asteroid means 'star-like' in Greek. It is believed that Jupiter's powerful gravitational influence and occasional near encounters with Mars can gravitationally perturb some of these asteroids from their conventional orbits. Like the effect on the comet Shoemaker-Levy 9, which was sent crashing into Jupiter in 1994, these huge mountains of rock can sometimes end up hurling toward the inner Solar System, posing a major threat to our own planet with potentiallydevastating consequences. In the most powerful telescopes, these mountain-sized objects appear little more than a starry speck of light moving among the background stars. Those that approach the Earth with relative closeness are referred to as near-Earth objects (NEOs) or near-Earth asteroids (NEAs) and are monitored closely by ground-based observers to ensure their orbits do not become critical. Whilst most of this work is carried out in the more densely populated Northern Hemisphere, a small but dedicated patrol is also undertaken here in Australia.

The spacecraft *Galileo* was the first to observe asteroids at close range during a fly-by through the asteroid belt in the early 1990s. The irregularly shaped rocky worlds encountered were Gaspra and Ida, looking somewhat like floating potatoes. Furthermore, and to everyone's surprise, *Galileo* also discovered that Ida indeed has its own tiny satellite, Dactyl, which is thought to be a possible fragment from a past collision.

## Observing asteroids

Moving among the fixed background stars we can view some of the major asteroids as they make their way across the night sky over several months. Ceres, Pallas, Vesta and Juno respectively are the largest among the few asteroids that can be detected using a pair of binoculars. Popular astronomy magazines and computer-based ephemeris software are the best reference sources for locating their positions in the night sky.

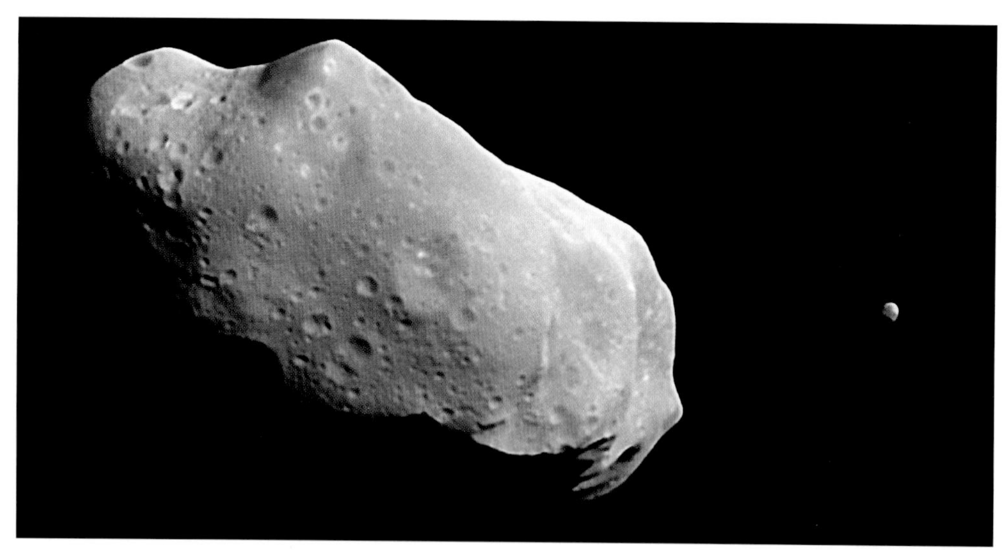

A striking image of asteroid Ida taken from the *Galileo* spacecraft on 28 August 1993. Ida is about 56 kilometres in length. To the surprise of mission scientists, Ida has a tiny orbiting satellite of its own, seen to its right. Later named Dactyl, it is about 1.5 kilometres in diameter. *(NASA/JPL)*

An asteroid's apparent brightness can vary for a number of reasons. One factor relates to its size and relative distance from Earth. Another relates to the different types of surface materials and distribution thereof which governs the most noticeable changes. Some asteroids have globally darker surfaces that reflect the Sun's light less efficiently, while others, such as Vesta and Eros, have much more efficient reflective properties.

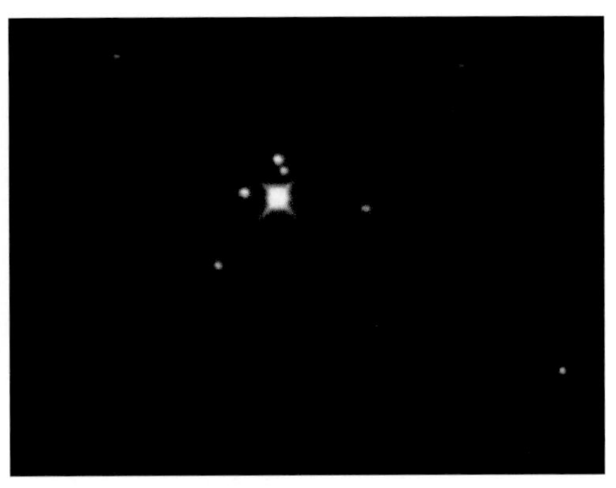

Bright asteroids can be seen noticeably moving among the stars in a matter of hours. Using a 250mm (10in) f/5.8 Netownian and 35mm Kodak VR 400 film, these two 5-minute exposures were taken 4 hours, 40 minutes apart to reveal the movement of Vesta on 17 April 1985. *(Courtesy Steve Quirk)*

# Comets

When discussing Solar System bodies with distances beyond Pluto, the numbers in terms of kilometres can become mind-boggling and require a more convenient numeric system. This is why astronomers came up with the *astronomical unit* (AU). 1 AU is based on the mean distance of Earth from the Sun, which is roughly 149.6 million kilometres. There are of course other measures, such as distance in light years and something known as a parsec; for the purposes of this book, however, we need not concern ourselves with them.

Far off in the outer regions of the Solar System, near the orbit of Neptune and extending far beyond Pluto to a distance of nearly 1000 AU, is a region known as the Edgeworth-Kuiper belt (most often referred to as the Kuiper belt). Similar to the asteroid belt, this frosty outer region is swarming with billions of small ice-dust worlds left over from the creation of the Solar System some 4.5 billion years ago. Kenneth Edgeworth hypothesised its existence in the late 1940s, while Gerard Kuiper later independently suggested it may be the origin of short-period comets (those taking less than 200 years to orbit the Sun) such as comet Halley, which returns to the inner Solar System every 76 years. Another region believed to be the source of long-period comets (comets that take many thousands of years to return) is called the Oort cloud.

Faster films allow for shorter exposure times in order to capture a portrait of a comet against a starry background. This 10-minute exposure of comet Levy was taken in September 1990 using ISO 1000 hypersensitised film through a 250mm (10in) f/5.8 Newtonian telescope. *(Courtesy Steve Quirk)*

This is said to be a Solar System-encompassing, spherical halo of comet nuclei, extending out to around 100 000 AU. With the same subtle effect that can influence the orbit of an asteroid, slight gravitational tugs from the planets can also alter the routine passage of a comet, occasionally drawing it into the inner Solar System.

Comets are believed to contain many of the organic materials essential to the origin of life. They can be likened to a compacted dirty snowball several kilometres in diameter made up of frozen gases, dust and water ice. As comets venture into the inner Solar System along their highly elliptical orbits, this material is heated and released into space, initially forming a *coma* or halo-like hood about the nucleus. On closer approach to the Sun a comet reacts to charged solar particles, forming a tail that always points away from the Sun. In some colour photographs this tail is often revealed as two distinct components—the brighter yellow dust tail and a blue tail of ionised molecules.

When the Earth passes through a remnant trail of comet debris we experience meteor showers as these fine particles burn up high in our atmosphere. Halley's comet is perhaps the most well known, having last returned in 1986. At that time it was greeted by the *Giotto* spacecraft which took pictures of it at close range. Comets Hale-Bopp and Hyakutake caused quite a sensation in the 1990s. Currently en route for an encounter with comet Wild 2, a space probe called *Stardust* is the first US mission launched to robotically obtain comet samples, and is scheduled to return the samples to Earth in 2006.

## Observing comets

Against the backdrop of a starry night sky, a comet's position can be seen to constantly change night after night. In fact, during a long photographic exposure while tracking the head, or coma, of a comet through a telescope or telephoto lens, background stars are recorded as streaks in the final picture and lay testament to the comet's motion. But this movement in terms of earthbound naked eye observation is very small. It still amazes me during bright comet apparitions that receive wide public attention, how many people claim to have seen it streaking across the night sky. This is a fanciful and misleading report from someone who has in fact not actually seen the comet. If they appear sincere then they have perhaps mistaken it for a satellite passing overhead or a sporadic meteor and should be delicately corrected by those in the know.

Many comets have been independently or co-discovered by astronomers examining photographic plates, while others have been detected from amateur's backyards, either visually or with a CCD camera. The initial appearance of most newly found comets is generally unimpressive, requiring the aid of a telescope

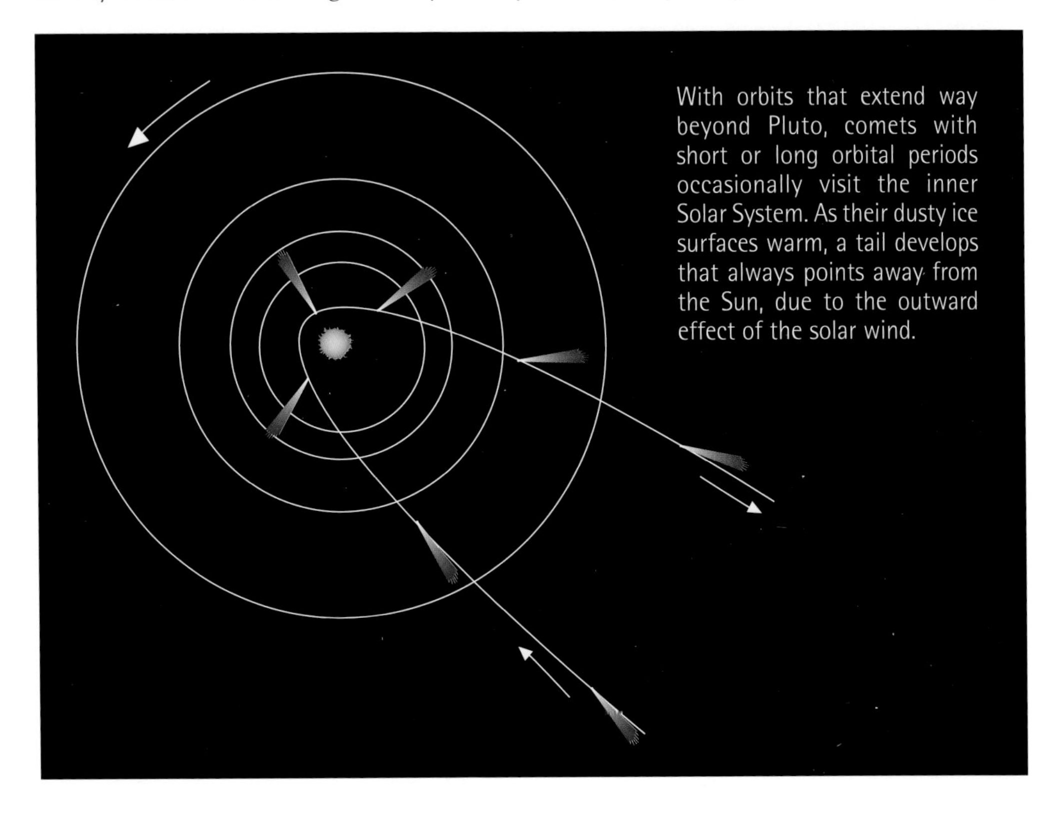

With orbits that extend way beyond Pluto, comets with short or long orbital periods occasionally visit the inner Solar System. As their dusty ice surfaces warm, a tail develops that always points away from the Sun, due to the outward effect of the solar wind.

Comets do not streak across the sky like a meteor but move slowly among the background stars. The movement of comet Hyakutake is evident in this 20-minute exposure taken on 23 March 1996. A 35mm SLR camera was used with a 300mm f/4 lens and Kodak ISO 400 film. *(Courtesy Steve Quirk)*

or moderately powered binoculars for positive identification. In the field of view they can appear small and faint, much like a fuzzy halo of light with a concentrated star-like centre. In most cases, these comets never achieve any great levels of visual brightness and public knowledge of their existence is virtually zero, but on rare occasions one becomes bright enough to be seen easily with the unaided eye.

Particularly in the case of faint comets, a good ephemeris providing coordinates for a particular target will enable you to hunt down this fuzzy vagabond among the stars. There are numerous wonderful software programs available today that provide up-to-the-minute database upgrades via the Internet. Comets appear at their best around nearest approach to the Sun (perihelion) and thus are usually best observed after sunset or before sunrise.

It seems that every time a bright naked-eye comet receives media attention, every man and his dog heads for the local camera store to buy a telescope. More often than not this leads to disappointment and eventually lots of second-hand telescopes ending up in the 'for sale' section of the classifieds. Since a comet and its tail can span several degrees of the sky, it is best viewed with binoculars or a wide-field, short focal length refractor. Common camera store beginner's telescopes produce a narrow field of view, thus limiting how much of the comet and tail can be seen, and also reducing its apparent brightness. Photographing a bright comet can be as simple as a short exposure of between 30 to 60 seconds using a 35mm SLR camera mounted on a sturdy tripod. With standard ISO 200–400 speed film and your camera set to largest aperture (i.e. lowest f number), nice results can be achieved if the comet is well above the horizon around late dusk or early twilight hours. Set the camera's focus to infinity and the shutter setting in the 'B' position for long-exposure control with a shutter-release cable. More impressive close-up shots can be achieved using a telephoto lens. To avoid smearing of stars or background landscapes from shutter release vibrations, use a piece of black cardboard placed over the camera's objective for a few seconds both before and after shutter release.

Amateur astronomers are very active comet discoverers and these days a number of Australians have their surnames attached to various comets. Aside from luck, finding a comet requires observing skills, determination and diligence for protracted periods of systematically scanning the skies. However, some discoverers are simply lucky and stumble across a new comet without actually seeking one. But even still, this requires knowledge of the sky and the aforementioned observing skills in order to discern its true nature. Who knows, perhaps your name may be immortalised some day.

## Meteors

Small rocky fragments known as meteoroids are derived either from asteroids or the debris trail left by a passing comet, and regularly fall victim to the protective atmosphere of the Earth. Burning up in a brief streak or flash of light high in the atmosphere they are then called meteors—a less romantic name for a 'falling star'. Occasionally a metallic or rocky fragment may reach the ground after breaking up high in the Earth's atmosphere in a bright flaming, and sometimes explosive, fireball. The remains are called meteorites. Since most of us are usually asleep or indoors during the dark hours, most of these exciting moments are missed.

While perhaps the most famous meteor impact site on Earth is in Arizona, Australia has Western Australia's Wolf Creek crater. It rises roughly 25 metres above the surrounding desert, while its crater floor lies some 50 metres below the highest peak on the rim. Impact glass created by the immense instantaneous heating and oxidised meteoric iron have been found at this site.

On just about any clear night you can look up with attentive scanning eyes and be fairly assured of glimpsing at least one moderately bright streak of light passing overhead within 30 to 60 minutes. Even if it appears momentarily from the corner of your eye you can be pretty certain you've witnessed the fiery death of a speck of celestial debris. In darker country skies you'll often see many more, since most are quite faint and get lost in the perpetual haze of suburban and city light pollution.

The terms 'falling star' and 'shooting star' are how most people identify these celestial streaks of light. What we are actually seeing are tiny grain-sized, sometimes stone-sized, particles entering our skies at extremely high speeds, which burn up at a height of around 90 to 100 kilometres in the atmosphere due to heating by friction. These random and unpredictable invaders of the night sky are known as sporadic meteors. Large, fiery meteors lasting more than a second or two are called fireballs and can often be seen breaking up into two or more pieces. They can leave long smoke trails in the sky, called *trains*, which on still nights can persist for long periods after the destruction of the meteor. When a large meteor enters the atmosphere and continues travelling through it, a pressure wave is created, known as an *infrasonic signal*. This pressure wave is similar to that created by an explosion and, depending on the size of the meteor and distance from the observer at the time it explodes, often results in an audible bang or sonic boom from the sky.

## Meteor showers

Each year meteor showers spray hundreds, even thousands of fiery streaks across the night sky, radiating from within the boundaries of several well-known constellations. If these were indeed falling stars, our chances of winning the lottery would be most favourable! But, keeping in mind the nature of a star, we wouldn't be around to enjoy the spoils of our good fortune.

The truth is that meteor showers, like the famous Leonid shower, are the result of short-period comets that circle the Sun every few years or decades. These frequent visitors are relatively easy to find and are routinely tracked by astronomers, while long-period comets on the other hand spend most of their time in the dark recesses of space far beyond Pluto. With little warning, one can swoop in from the outer Solar System and pass uncomfortably close to our planet. Along with the Leonids, the visibility times for other meteor showers, such as the Orionids and Geminids, can be predicted, including when and where they can best be seen.

The magnificent Leonid meteor shower on 18 November 2001. *(Courtesy Steve Quirk)*

So why do meteor showers occur? As previously mentioned, comets react to the effects of the solar wind and increased surface temperature, thus forming a tail of fine particles and fluorescing gas ejected from its surface millions of kilometres into space. The Earth and other planets continually sweep up this scattered debris along their orbits around the Sun. When the Earth occasionally intersects the orbital path of a body such as comet Tempel-Tuttle we encounter a much more densely populated region of dust particles. The temporary result is a gradually intensified stream of micro-meteors, burning up as they bombard the upper atmosphere. After reaching a peak at some point, activity begins to wane. Comet Tempel-Tuttle returns once every 33 years and is the source of ejected material responsible for the Leonid meteor shower. Meteor storms or showers are named after the constellation they appear to radiate from. The point from which the streaks of light appear to stream away in all directions is called the *radiant*.

Astronomers record the number of meteors seen each year using long-exposure photography, video and other, more sophisticated computer-based tools. The number of meteors observed during a shower changes each year in relation to the recentness of the comet's passage around the Sun and the density of debris with which the Earth intersects at a given time. Some years, the number encountered may be quite disappointing while at other times it can be overwhelming. Also, depending on your geographical location and the predicted time of maximum intensity, observing the event may not always be favourable. Even if maximum can be observed above the horizon, but the radiant is low in the sky, then many fainter meteors are difficult to detect due to the scattering of this faint light through a denser and less transparent atmosphere.

Managed by dedicated amateur observing groups around the world, a system of averaging meteor rates (reported by observers at different locations and applying correction factors to each) produces a standardised view if observed under a 'perfect' sky. The resulting estimated rate is referred to as ZHR, or *zenithal hourly rate*. When a shower is intense it can be almost impossible to visibly count the numerous flashes and streaks that can occur in a short space of time. Today many amateurs apply modern and highly light-sensitive video-based CCD cameras, coupled with computers and special automated detection software, to make this task much easier, yielding much higher ZHRs than ever before.

In 2001 the Leonids were well situated for the Australia–Asia regions to view them, yielding a ZHR of around 3500 meteors. In 2002 observers in Australia essentially missed the best of the event, since the radiant rose in Sydney about 6 hours after the peak. At that time the ZHR was below 200, and with a full moon present and the initial low elevation there was little observable activity. It will be some years before the Leonids will produce such spectacular results as were seen in 2001.

A meteor shower is a spectacular event to watch and most certainly worthy of a dedicated trip to a dark-sky location just to observe or even photograph.

| **Annual meteor showers** | **Month** | **Constellation** |
| --- | --- | --- |
| Lyrids | April | Lyra |
| Eta Aquarids | May | Aquarius |
| Delta Aquarids | July | Aquarius |
| Perseids | August | Perseus |
| Orionids | October | Orion |
| Leonids | November | Leo |
| Geminids | December | Gemini |

# Tools for observing

I think it would be fair to say, at least in a large number of situations, that most people (particularly kids) tend to jump in at the deep end when engaging a new gadget—especially if it involves set-up instructions. Often ignoring or discarding boring manufacturer instructions or advice, we later find ourselves having to refer back to them because our exciting new purchase or gift didn't quite perform as expected. If you're a true beginner then it's a good idea to familiarise yourself with basic telescope designs, performance characteristics and associated accessories so you'll be well equipped to make an educated choice when it comes to using, upgrading or purchasing your first telescope.

A quality telescope will reward you with a lifetime of enjoyable observing. With the variety of telescopes available today, design and performance characteristics must be considered to ensure you select an instrument that meets with your expectations and your budget. Some questions you may ask include: Can I use it for taking photographs? Will I clearly see the divisions in the rings of Saturn? Do I need to transport it a lot? Will it fit in the car? What accessories will I need? Does it depend on a power supply? If you are planning on buying or upgrading a telescope then these are some of the questions you should ask yourself and ultimately your optical dealer. Although some telescopes and their supplied mounts are better suited to certain tasks than others, your first consideration should be an instrument that can deliver the sharpest view possible, with good image contrast and overall system stability.

## Telescope basics

The most obvious function of a telescope is to magnify distant objects. When observing faint and/or distant celestial targets, two equally important requirements come into play—the ability to amplify faint light and the ability to resolve fine details, commonly referred to as the telescope's resolving power. This is perhaps the most important yet disappointing factor for most first-time small-telescope users. For example, if viewing Saturn with a 60mm (2.4in) telescope and an eyepiece producing a magnification of 30X power, we see a reasonably bright yet diminutively small ringed world. Human nature demands that we must now zoom in for a closer look. But with each step of increased eyepiece magnification, our subject becomes progressively dimmer until we have exceeded the telescope's practical light-gathering limit. A telescope can be thought of as a sort of light funnel and from this example we can say the bigger, the better. Like the aperture rating of a camera, both light-gathering ability and resolving power of a telescope are governed by the diameter of the objective lens or primary mirror.

Other specifications you will encounter are focal length and f/ratio. A telescope's focal length is defined as the distance from the centre of a lens or curved mirror to its point of focus. The ratio between the focal length and its effective aperture is known as its f/ratio (e.g. f/10 or f/5). Along with the telescope's aperture, manufacturers may quote either the focal length or f/ratio. Telescopes with long focal lengths are preferred for planetary work because longer focal lengths produce larger image scales.

The following guide is aimed at familiarising the beginner with three commonly available telescopes and their basic differences. The Internet and astronomy periodicals found at most news stands are excellent resources for more product-specific information. Of course, the best way to make an educated choice is to try before you buy if at all possible. Amateur astronomy clubs provide a good starting point and you may be pleasantly surprised how eager others are to demonstrate and discuss their equipment and experiences.

## Refractors

When most people visualise a telescope in their mind's eye it's usually the refractor type. They are comprised of a curved (convex) objective lens housed at one end of a long sealed tube. The eyepiece is located at the opposite end and is coupled to a smaller sliding tube that is focused via a simple rack-and-pinion knob. There are other designs on the market today where focus is achieved with an adjustable primary objective lens cell that is hinged to a long threaded rod and single focus knob at the rear of the instrument.

Classic entry-level 60 x 700 telescopes can be found in most camera shops and department stores. The '60' represents the objective lens aperture in millimetres while the '700' represents the focal length in millimetres. It is generally held that these telescopes should be avoided, and for serious observing in particular, this is quite true. However, if this is all that is available to you for now then just be sure to keep your expectations to a minimum. A telescope of this specification can produce nice simple views of the Moon and very modest

views of the planets. For example, one can expect to see Jupiter and hints of its encircling cloud belts. Its four major moons, which change positions each night, are also easily spotted. Saturn and its rings are also easily seen at medium to high powers including its largest moon, Titan. You can witness the changing phases of Venus, too, and in extremely good viewing conditions glimpse the polar ice caps and subtle dusky surface markings of Mars during times of closest approach with Earth. Despite the criticism you may hear from serious amateurs about these telescopes, most of them began their lifelong hobby with just such an instrument.

Larger aperture, high-performance achromatic or apochromatic refracting telescopes are ideal for observing the planets, but are proportionally more expensive per centimetre of aperture when compared with more cost-effective reflecting telescopes.

Refracting telescopes suffer from varying degrees of an optical defect known as *chromatic aberration*. Chromatic aberration is the failure to bring light of all wavelengths to a common point of focus. In other words, blues and reds appear slightly shifted. To correct the problem, modern day refractors are comprised of two objective lenses coupled together and these are called *achromatic refractors*. But even achromatic (abbreviated to achromat) optics still exhibit some amount of chromatic aberration. When looking at bright stars, the limb of the Moon or planets such as Jupiter or Venus in particular, the effect is a purplish haze or halo

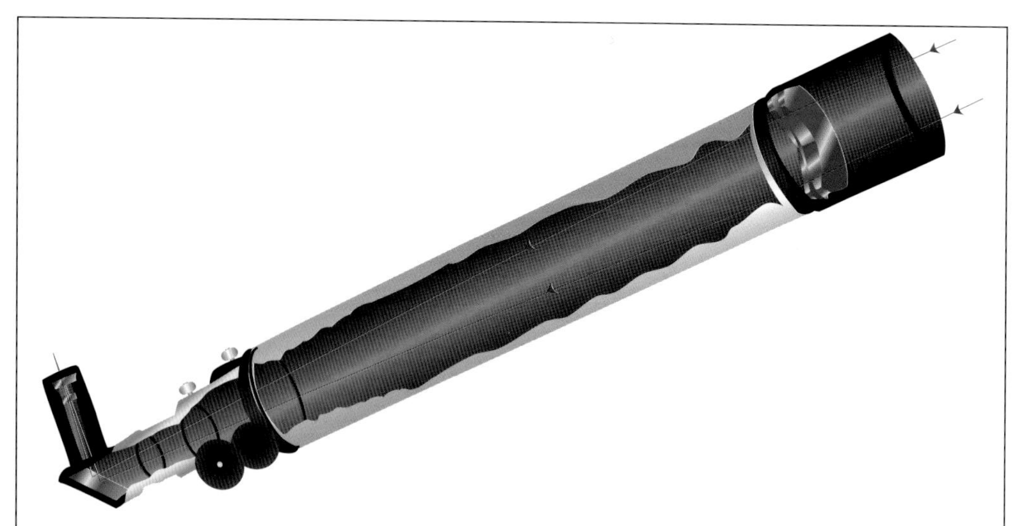

Light passes through a primary objective lens at the front of the refractor telescope, bending the incoming light down a tube. This focused point of light is then magnified via a series of smaller lenses at the end of the tube (the elements within an eyepiece). While a refractor can deliver high-quality images, the design is expensive at large apertures.

around the subject. The most efficient refractor design is called an *apochromatic refractor* which uses three coupled lenses of crown and flint glass, greatly minimising the effect to almost negligible levels, thus producing much sharper, contrast-rich images. But apochromatic refractors are extremely expensive, particularly in large apertures of 127mm (5in) and greater, and usually beyond the practical budget limit for most of us. There is, however, a makeshift solution for more commonly purchased achromatic telescopes in the form of a filter that can substantially reduce this purplish halo. Called minus violet filters, they simply screw into the barrel of an eyepiece like any other filter and reduce visible chromatic aberration without greatly altering the natural appearance of the subject. They do not convert an achromat to an apochromat.

Another symptom seen occasionally in refracting telescopes is known as *spherical aberration* and is the result of poorly crafted optics. In short, the resulting effect is an inability to bring a subject into sharp focus. For planetary work, the advantages of refracting telescopes are primarily the sealed optical tube with no central obstruction and longer primary focal lengths. While common apertures of 60mm (2.4in) to 150mm (6in) are available, apertures of 100mm (3.9in) or greater are best suited to detailed planetary observations. Refractors are also susceptible to dew settling on the primary objective.

## Newtonian reflectors

Newtonian reflectors are popular for a number of good reasons. They not only rival the performance of other optical systems but are extremely economical per centimetre of aperture.

Newtonian reflectors are comprised of two mirrors. The larger primary mirror is parabolic and reflects incoming light rays to a smaller flat secondary mirror mounted diagonally. This secondary mirror diverts the light to an eyepiece near the skyward end of the telescope, where the image is subsequently brought into focus with a sliding tube rack-and-pinion arrangement. Since they reflect light, Newtonian reflector telescopes do not suffer from the chromatic aberration found in refracting instruments.

While often less portable in larger designs, Newtonian reflecting telescopes offer the most aperture per dollar. Here, a camera is mounted piggyback for tracked wide-field photography.

Like the department store refractor, there are also low-grade introductory reflectors floating about. Most specialist astronomical telescope dealers can supply you with an excellent beginner's 110mm (4.5in) or 150mm (6in) reflector that can produce outstanding views. The latter is considered the minimum aperture suitable for useful observations. Commonly available Newtonians range in focal ratios from f/4.5, f/5 to f/6 and some up to f/8. Apertures typically range from 110mm (4.5in) to 410mm (16in). Instruments with apertures 200mm (8in) and larger are capable of producing beautiful close-up and detailed views of the Moon and the planets.

Kits are available for those wanting to construct their own instruments and many instructional books are available. While equatorial mounted systems are

Light passes down a Newtonian telescope's tube to a parabolic primary mirror at the bottom. It is then reflected forward to a secondary mirror and out the focusing tube, both of which are located near the front of the telescope.

the desirable option for protracted lunar and planetary observations, another popular design is the Dobsonian system. This is a Newtonian optical tube assembly mounted on a basic squat-style altazimuth platform usually made of wood. At high powers these systems require constant and annoying recentring of the target on both axes. They are best suited to casual deep-sky observing.

A minor side-effect of the otherwise overwhelming benefits of a reflecting telescope (particularly in close-up planetary work) is the open tube construction, which is more susceptible to local convection air currents that can randomly distort the incoming light in all manner of ways. However, by selecting your observing site carefully (such as a grassy area) and allowing a few hours for the mirror and tube to acclimatise to the surrounding air, much of this effect will be reduced. Unlike refractors, reflecting telescopes have a central obstruction due to the secondary mirror. This results in a slight loss of image contrast. In an equatorial system, you also need to rotate the optical tube for convenient positioning of the eyepiece when viewing different parts of the sky.

## Catadioptric (Cassegrain) designs

There are two commonly available catadioptric designs, the Maksutov-Cassegrain and the Schmidt-Cassegrain. Both are extremely portable telescopes based around a mirror–lens design involving greater manufacturing costs compared with a Newtonian, yet their versatility makes them an appealing option. Unlike the conventional rack-and-pinion focusing system used in most refractors and Newtonian telescopes, catadioptric scopes are focused by the inward–outward adjustment of the primary mirror via a focus knob attached to an internally threaded rod.

While refractors have the advantage over reflecting telescopes of an unobstructed primary objective, Matsukov- and Schmidt-Cassegrains have an inherently larger secondary central obstruction compared with Newtonians. They do, however, share the common advantages of a sealed tube and longer primary focal lengths, with ratios of f/10 to f/15 being the most common. Like refractors, both these instruments are also susceptible to dew settling on the primary lens; dew covers are a favourable option.

These are excellent all-round instruments offering the best of both worlds. Many are supplied today with smart GOTO electronics on an altazimuth driven mount which can further be mounted to an optional equatorial wedge for more accurate astrophotography.

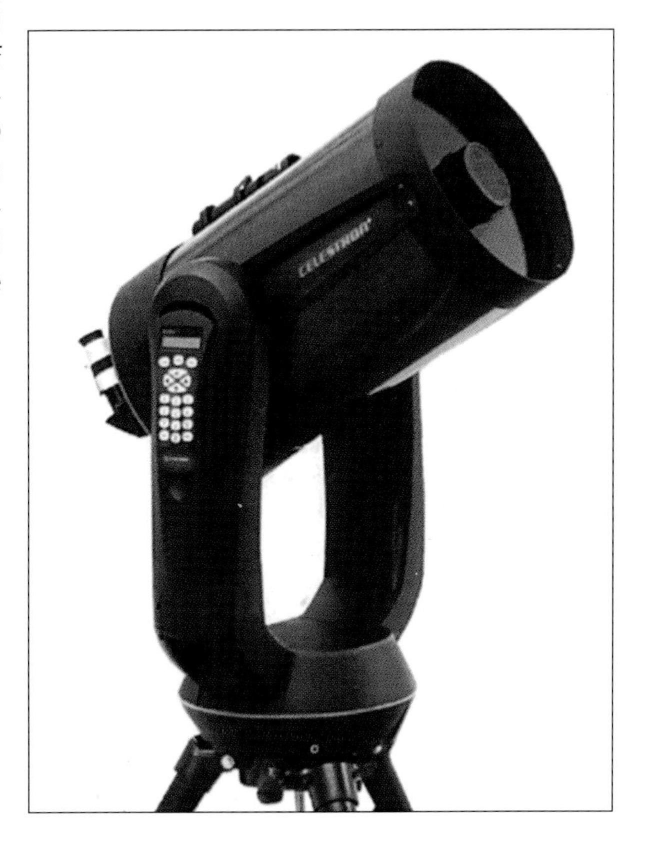


Light enters a Schmidt-Cassegrain telescope through a thin, sealed corrector plate at the front of the telescope. Upon reaching the back of the telescope, it is reflected forward by the primary mirror to a convex secondary mirror. The secondary then redirects this light back down the tube through a hole in the centre of the primary mirror to the focusing tube and eyepiece. This design lends itself to greater portability, but such telescopes are generally more expensive than Newtonian designs. *(Courtesy Tasco Australia)*

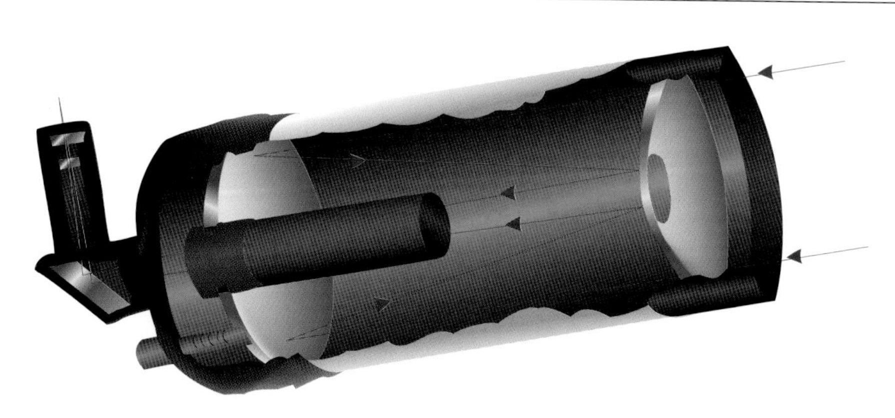

Similar to a Schmidt-Cassegrain, a Maksutov is also a sealed reflecting telescope. Light enters via a spherical corrector plate (convex side towards the primary mirror). This light is then reflected back from the primary mirror to a small aluminised spot on the inside centre of the corrector plate. The light is then focused down a tube, in the centre of the primary mirror, to the eyepiece.

## The focuser

When observing objects at extremely high powers, smooth focusing is absolutely essential. The focuser's rack-and-pinion assembly is yet another area some manufacturers have historically cut back on in terms of quality. Plastic units are often the worst, but all-metal assemblies can be just as sloppy to operate if gear and tube tolerances are too great. A typical symptom of a poor focuser assembly is large shifting in the image when making focus adjustments. Depending on the magnification being used, the

A smoothly adjusting focuser is important, especially when observing at higher powers; a poor focuser will shift dramatically. This standard SkyWatcher focuser takes both 50mm (2in) and 31.7mm (1.25in) eyepieces.

worst-case scenario is an image that is completely removed from the field of view when adjustments are made.

Most common focusers take 31.7mm (1.25in) barrel eyepieces while others take 50mm (2in) barrel eyepieces with a 31.7mm (1.25in) eyepiece adaptor. Less commonly found these days are focusers that take only 24.5mm (0.965in) barrel eyepieces. The range of quality eyepieces available in this size is virtually zero.

## Aligning the finderscope

The purpose of a finderscope is no different from that of a shooter's riflescope, only in astronomical applications those supplied with most common models project an upside-down image. This is simply a case of manufacture and supply cost-saving and it is of little significance when observing the night sky. The common finderscope is usually a low-power refracting telescope in miniature but with cross hairs for centring a target. For those who prefer the a non-optically altered naked-eye view of the sky when aligning an instrument, the reflex finder projects a tiny red dot onto a transparent glass or plastic surface from a small battery operated LED.

No matter which type your telescope has, the important thing is it must first be aligned with the optics of the main telescope to ensure you enjoy hassle-free observing. To do this, first insert a low-power eyepiece into the telescope's focuser. Choosing an easy target like the Moon or a tree on a distant hilltop, move the telescope about until the object is in the centre of the field of view. Now look into the finderscope and adjust the knurled screws until the same target is

Easy targeting of a telescope makes for a more productive observing session. By centring the main telescope on an easy reference object with a low-power eyepiece, the finderscope can then be adjusted via the pivoting screws so that its internal cross hairs intersect with the same target.

centred in the cross hairs. With this rough alignment in place, you can then refine targeting accuracy using a bright star. Back at the eyepiece, adjust the telescope so the star appears dead centre, then make fine adjustments at the finderscope to ensure the star is centred in the cross hairs. Though not essential, this task is best carried out while a telescope is tracking the sky (drive motor enabled). While bigger is considered better in terms of finderscope aperture for deep-sky observers, it is not an essential requirement for locating the brighter planets and the Moon.

## Collimation

Just like the importance of a finely tuned television antenna for receiving the clearest picture on a television screen, a telescope requires a finely tuned light path to produce the sharpest image in the eyepiece. It is important to ensure that your telescope's optics are properly aligned; this is referred to as *collimation*. A poorly collimated instrument will produce inferior views. The handbook with your telescope should describe checks and adjustments in more detail.

In a Newtonian design, a good instrument will have at least three adjustment screws located at the rear of the instrument for movement of the primary mirror holder and three associated locking nuts or screws to secure each adjustment. The secondary mirror should also have three pivot adjustment screws. These features will vary slightly in different manufacturer designs but the principles for adjustment are the same. In Schmidt-Cassegrain telescopes, collimation is achieved by adjusting screws located on the exterior of the secondary mirror at the front of the instrument. A new telescope straight out of the box may not be correctly collimated due to the stresses of shipping from the manufacturer to the dealer and to your home. You can do a simple visual check and adjustment in accordance with the manual provided, but more accurate collimation can be achieved using a variety of collimating tools that are available. These devices are placed in the focuser and can be as simple and affordable as a cap with a tiny pinhole at its centre or a steel tube with a pinhole and cross hairs. A more costly device is a sophisticated laser collimator.

Critical collimation is always preferable however not as demanding in reflectors with focal ratios of f/6 or greater. Instruments with focal ratios of f/5 or f/4.5, on the other hand, have steeper light cones and consequently require more diligent optical alignment to avoid possible distortions in fine planetary detail. Once you have collimated your telescope a few times it will become second nature, so don't be afraid to give it a try if you are sure you understand the principles fully.

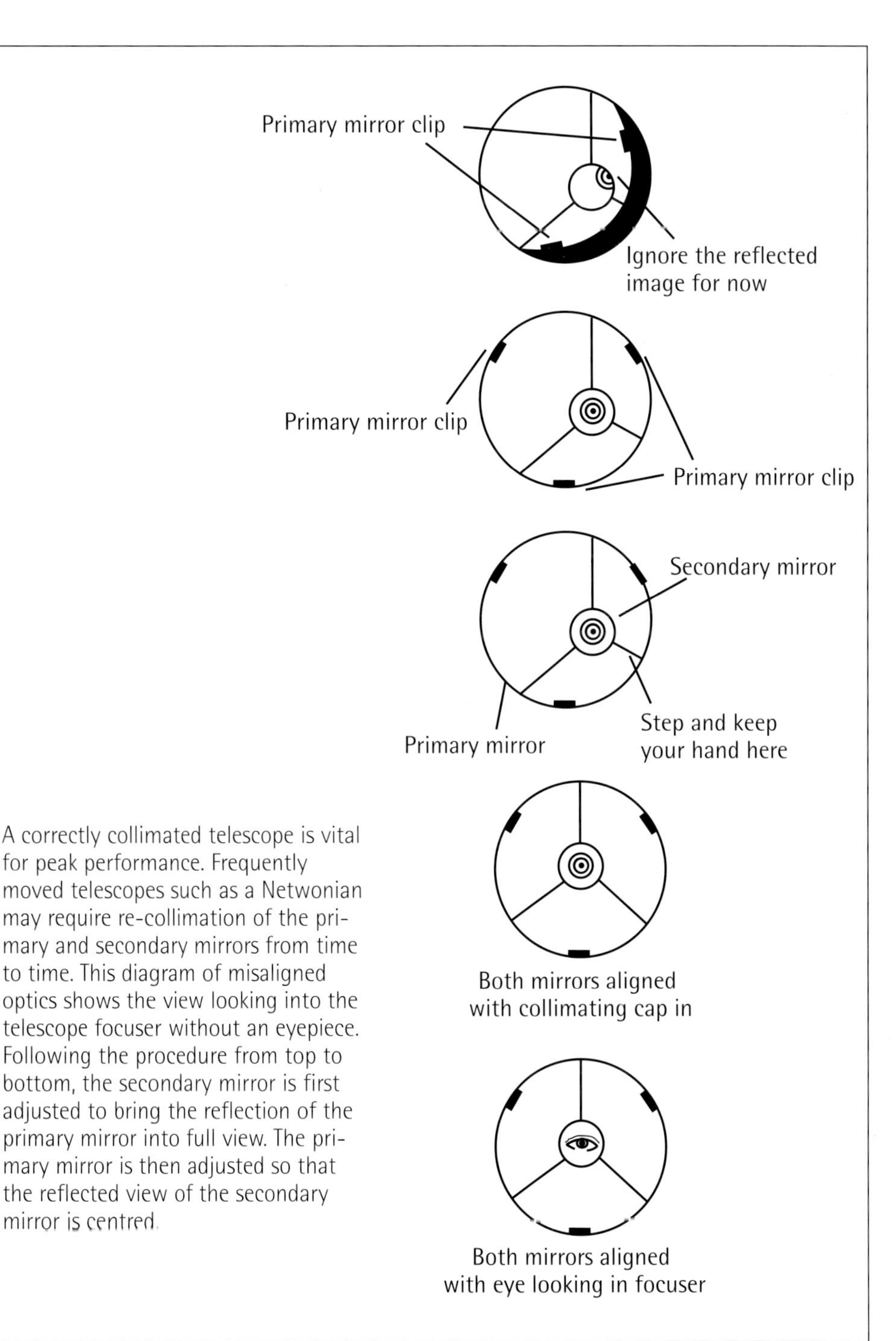


A correctly collimated telescope is vital for peak performance. Frequently moved telescopes such as a Netwonian may require re-collimation of the primary and secondary mirrors from time to time. This diagram of misaligned optics shows the view looking into the telescope focuser without an eyepiece. Following the procedure from top to bottom, the secondary mirror is first adjusted to bring the reflection of the primary mirror into full view. The primary mirror is then adjusted so that the reflected view of the secondary mirror is centred.

# Telescope mount and axis drives

An area that in years gone by had been seriously neglected by manufacturers was the supplied telescope mounting. Your telescope's mount is the axial hub between the tripod or pier on which the optical tube assembly is supported. Combined, the tripod or pier and axial mount form the mounting system. Due to consumer feedback, much effort was initially put into improving optical performance, and for the most part a large number of mass-produced optics today far exceed the quality of their predecessors of the 1960s and 1970s. A fine optical instrument on a wobbly mount or tripod can be just as frustrating as an image that can't be focused properly. Some manufacturers continue to supply telescopes today with inadequate mounts in order to keep their costs down, so you should be careful to choose an instrument provided with a solid mounting system.

There are three common styles of telescope mount you will encounter: the conventional altazimuth mount with up–down, left–right movement; the computerised GOTO altazimuth mount with optional equatorial wedges; and the German equatorial-style mounts that move in right ascension and declination motion. First and foremost, the mount must be of solid construction in order to reduce the amount of shakiness one can see in the eyepiece, especially at higher powers. A quality mounting system will have quick vibration-dampening characteristics—in other words, a subtle knock to the instrument will settle within 1 or 2 seconds. When adjusting the focus, slight movements of the focuser can result in minor tremors. This is particularly noticeable at high powers. In a solid mount that is rated to handle weights much greater than the total weight of the optical tube assembly, these vibrations should settle almost immediately. Tasco Australia distributed a well-manufactured budget 150mm (6in) telescope, the OP500 series, under their own OPTEX brand that

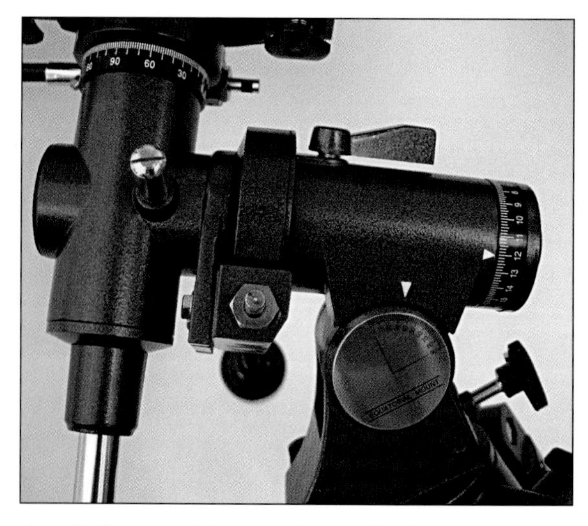

A typical equatorial mount for small telescopes. An unstable or payload underrated mount can prove to be very frustrating—equatorial mounts should be given serious consideration when choosing a telescope. Optional motor drives are generally available for external attachment to the worm gears on each axis.

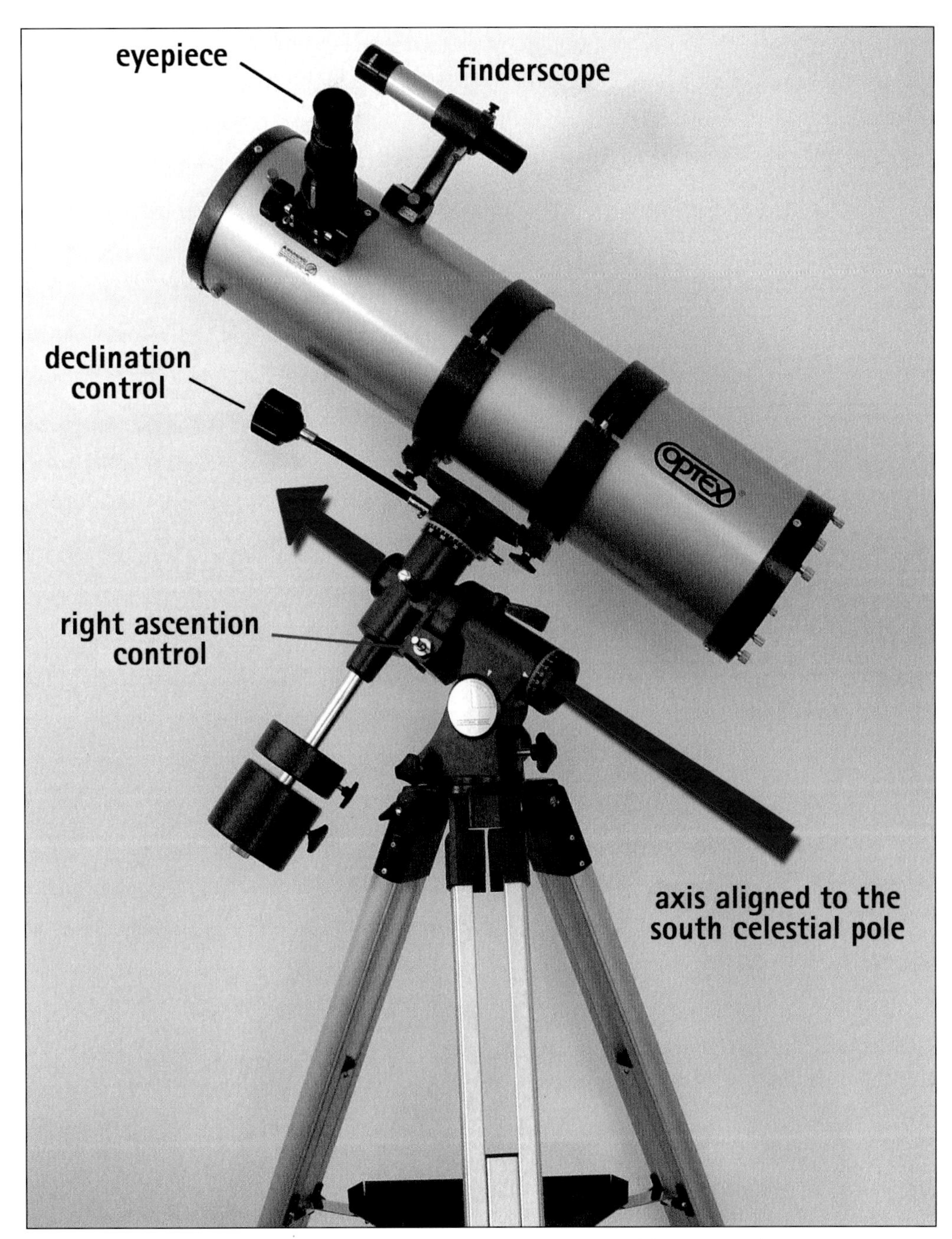


A cost-effective 150mm (6in) Newtonian reflector on an equatorial mount and tripod. The arrow through the polar axis indicates the direction to the south celestial pole.

was supplied with a very good German equatorial mount. This system was a giant leap forward in mounts supplied for the beginner to intermediate market that also provided excellent right ascension motor drive tracking.

Many supplied telescope mounts are on tripods and in some instances these can present a problem when viewing targets directly overhead; the optical tube or optional drive motors may encounter one of the tripod legs, making it impossible to reach your target without resorting to turning the whole tripod around and losing polar alignment. A pier-mounted equatorial system is the better option if available. It is wise to check the design of the tripod mount system for potential positional obstructions before buying it.

GOTO mounts are commonly altazimuth designs with encoders on each axis that feed back coordinates to a hand controller or a computer so targets can be located at the push of a menu button or enter key! These GOTO mounts can be mounted for equatorial tracking, using an optional equatorial wedge. This is the best solution for smoothly guided astrophotography and planetary imaging. Today, a large number of telescopes supplied with a sturdy equatorial mount now also incorporate GOTO functionality with fast slewing options.

The power of GOTO technology at the press of a button! Most systems today have in-built tours of the sky at any given time or location with commentary readout on the LCD display. Included is a huge preprogramed database of celestial objects and the ability to control the telescope from a personal computer.

A compact GOTO refracting telescope can be set up for touring the sky from any location within minutes.

The right ascension setting circle is used for finding celestial targets at a given east-to-west position in the sky. Like a 24-hour clock it is incremented in hours and minutes. At the rear is an adaptor for mounting a handy polar alignment scope.

The declination setting circle measures a telescope's movement in degrees north or south of the celestial equator. Once the telescope is correctly polar aligned, a star of known coordinates can be centred in the eyepiece and the declination ring adjusted to match that listed for the star. Once set, no further adjustment should be required unless the telescope is moved.

Another important, though not critical, feature in GOTO telescopes is quality 'setting circles' fixed on each axis of the mount. These allow you to manually find objects quickly when using coordinates given in an ephemeris. They are particularly convenient for tracking down planets during the daytime or fainter evening targets such Uranus, Neptune or an asteroid at night.

Equatorial and altazimuth mounts will have fixing hand screws for each axis. While loosened they allow for quick slewing of the instrument by hand until a new target is within the finderscope. Once located, tighten these screws to avoid the telescope drifting, then make further adjustments using the fine adjust knobs, screws or hand controller depending on your instrument.

A latitude scale, at the base of an equatorial mount, is used to set the mount's tilt angle in accordance with the observer's latitude north or south of the equator. For the Southern Hemisphere this corresponds to the height or elevation of the south celestial pole above the horizon (e.g. Sydney is at latitude 33.5°S, therefore you would adjust elevation to between 33 and 34 on the latitude scale).

## Drive motors

For protracted viewing enjoyment and astro-imaging it is best to have a mount that has in-built axis drives or allows for the attachment of optional drive units. Most important of all is the right ascension drive, so that a telescope is synchronised to the Earth's rotation, thus following and keeping targets in the field of view. Most modern drive systems offer the ability to advance or retard this motion with fine speed adjustments. This is useful, since there will be slight tolerance variables in the mount, drive gears and electronics, thus making it less than 100 per cent perfect. If your telescope mount is levelled and polar aligned with a great deal of accuracy, then little or no adjustment to the declination axis will be required. The addition of a declination drive is, however, very useful, particularly if you undertake rough polar alignment as I often do. Furthermore, it is useful for directional changes when observing and imaging the Moon at higher powers.

An optional RA motor drive fitted to an equatorial mount for keeping celestial targets centred in the eyepiece for protracted periods. The slew adjustment hand controller plugs into the modular connector at lower left.

An optional declination motor drive is fitted to an equatorial mount. Loosening the thumb wheel at left releases the gear drive for manual slewing of the telescope.

# Balancing a telescope

Correct balancing of your telescope on an equatorial mount, or in some cases a Schmidt-Cassegrain on an altazimuth mount, is important for accurate guiding and minimial strain on drive motors. When mounting additional equipment such as guide scopes and cameras, counterweights need to be adjusted or added to ensure the telescope doesn't flip over at certain viewing positions. Poor tracking in a properly polar aligned telescope is often the result of a poorly balanced instrument, causing drive clutch slippage. Your telescope manual will detail the procedure for doing this, so take the time to familiarise yourself with the procedure and avoid frustration later.

The right ascension motion of a Newtonian telescope on an equatorial mount. In the horizontal position seen here, the counterweights must be moved along the shaft to match the opposing weight of the optical tube assembly so that the telescope does not flip over. The optical tube must also be adjusted forward or backward within the mounting rings for balance in the declination axis. When mounting cameras or other accessories, this additional weight should also be taken into consideration during the balancing procedure. Balancing should be done with the axis screws loosened.

# The eyepiece

When it comes to eyepieces (oculars) for your telescope there is one simple rule to follow—buy only the best you can reasonably afford. Even if your telescope is of modest quality, a good eyepiece will at least project the best view it can deliver. At the other end of the scale, don't devalue your viewing experience with inferior budget eyepieces if you've already invested top dollar on a quality instrument.

There are a few primary factors that distinguish a good eyepiece from a poor one, as well as its suitability for a certain task. Three of these are *eye relief*, *exit pupil* and

A quality set of eyepieces will make your observing experience all the more rewarding and enjoyable. Purchase the best you can afford to ensure your telescope can deliver its greatest potential.

*apparent viewing angle*. Eye relief relates to the minimum distance required to place your eye up to the eyepiece and gain the maximum available field of view. Some economy high-power eyepieces have an eye relief sometimes less than one third of the ocular's focal length and should be avoided.

The point at which light rays from an eyepiece converge into the smallest focused diameter (like the pointed tip of a cone) is referred to as the exit pupil. This is where our eye receives the magnified image from the telescope's objective. The diameter at this point should fall within the maximum possible aperture of your pupil to ensure that your eye registers all the available light. As a guide, the maximum pupil aperture reaches approximately 7mm (0.27in) when completely dark-adapted but can be less in older people.

Wide-angle specifications have to do with the amount of edge-to-edge viewing angle, or apparent viewing angle, afforded by the eyepiece. A wide-angle eyepiece of, say, 40° or more will unfold a view with a greater sense of being out among the stars. In the case of a narrow-angle eyepiece the view can be almost tunnel-like, with a smaller defined circular boundary seeming to cut off the surrounding sky. The difference could be compared to looking through a ship's porthole with your face up against the glass then stepping back about a metre from the view.

There are a number of eyepieces available today of varying optical design, benefits and of course price tags, but in order to keep things simple let's discuss the more common types you'll most likely encounter. Orthoscopic eyepieces generally produce good contrast and sharp images across most of the magnifications available, however they typically produce a narrower field of view. Kellner eyepieces suffer somewhat from internal reflections when observing bright targets such as the planets or the Moon. This can produce an annoying ghostly-looking image, particularly where the planets are concerned.

As the saying goes, you get what you pay for, and that pretty well sums up the more expensive Plössel eyepieces. With excellent colour correction, a wider, flat field of view and good eye relief they are well worth the investment. At the top end are the magnificent 6- to 7-element Nagler eyepieces by Tele Vue Optics in the United States. While they are considerably more expensive than most mass-market eyepieces, they do offer superb eye relief, image contrast and wide-angle viewing. However, there are many mid-priced quality brands, such as Vixen, Meade and Celestron, with very good performance characteristics.

## Multicoated lenses and mirrors

This is a term you will encounter when checking the specification of refractors and eyepieces in particular. Optical coatings prevent reflection and scattering of light, which minimises light loss, thereby offering improved image contrast. An uncoated glass surface can lose up to 5 per cent of light transmission due to reflection and scattering. A single layer of anti-reflection coating can reduce loss to about 1.5 per cent. Multiple layers of different anti-reflection coatings can further reduce loss to as low as 0.25 per cent. The mirrors in most manufacturer reflectors are multicoated with silicon dioxide, which is especially needed in open-ended Newtonians, since the mirror is exposed to the elements and deterioration of the reflective layer reduces the performance of a telescope. Multicoating provides a higher level of light transmission and image contrast.

## The Barlow lens

Good quality Barlow lenses are useful tools for planetary and lunar work. Sometimes referred to as telenegative lenses, they effectively increase the focal length of any eyepiece by a factor of 1.5X, 2X or 3X. Therefore, if you have only one eyepiece, let's say 12mm (0.47in), inserting this into a 2X Barlow lens will produce a magnification equivalent to using a 6mm (0.24in) eyepiece. Since a 12mm eyepiece will generally have better eye relief over a 6mm eyepiece of the same type, a Barlow with the 12mm will maintain more pleasurable viewing.

A Barlow or telenegative lens increases the power of any inserted eyepiece by effectively increasing the focal length of a telescope. They typically come in 1.5X, 2X and 3X and are especially useful in photography of the Moon and the planets as an alternative to conventional eyepiece projection.

A word on using excessive magnification: beyond a certain point, increasing optical power will not improve visual detail and, depending on atmospheric seeing, the opposite may well be the case. The secret to increased image detail is aperture and steady skies.

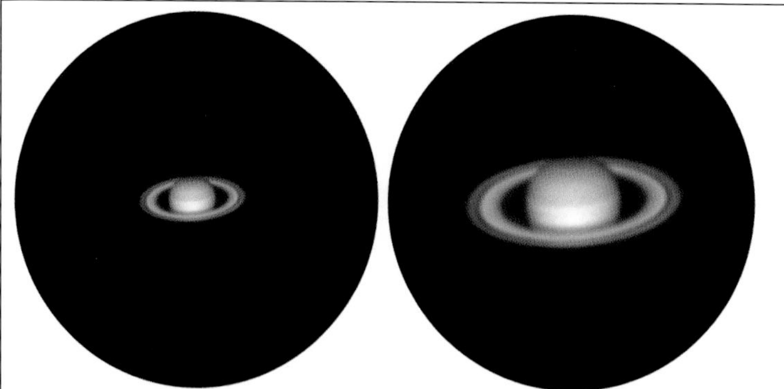

The image at right represents the eyepiece used in the left view combined with a 2X Barlow lens, magnifying the image of Saturn for closer inspection.

## Resolution

Resolution can be defined in terms of how much detail a telescope can see. Assuming that the optical system is correctly collimated, resolution is dependent upon the size of the rated aperture and the quality of the optical surfaces. Resolution is generally stated in arc seconds and can be determined by an empirical formula known as the Dawes limit, a method used to measure resolving power when splitting two very close stars of around 6th magnitude.

When we look at a star, we are seeing an infinite point source since it is so far away. Due to diffraction, we don't even see a point of light with a telescope but a circle of light called the *airy disc*. The arc second diameter of this disc decreases as a telescope's aperture increases. The airy disc is surrounded by increasingly faint concentric light rings and combined they make up the whole image called a *diffraction image*. The ability to separate two close stars, like a binary pair (two distant suns which orbit one another), is actually the ability to separate their airy discs. A telescope's ability to optically split closely spaced binary stars can be calculated by dividing 4.56 by the aperture of your telescope in inches (or 116 divided by the aperture in millimetres). Thus a 10-inch (250mm) telescope has a theoretical resolution of 0.456 arc seconds. The smaller the number the finer the detail that can be observed or photographed.

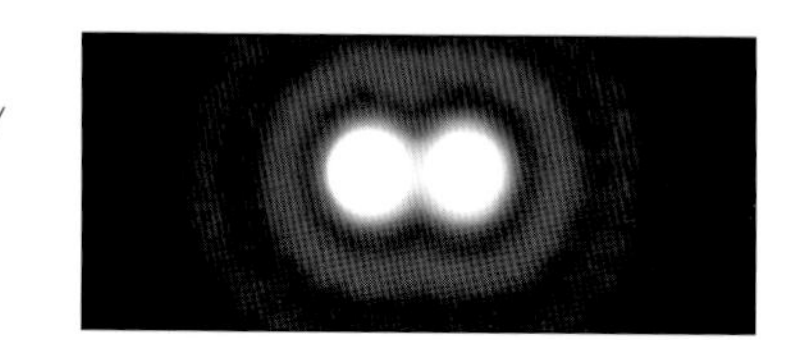

The resolving power of a telescope can be measured by its ability to separate two very closely positioned stars. *(Courtesy Tasco Australia)*

# Observing with binoculars

Binoculars are a great tool for observing bright comets and tracking down faint targets such as Uranus and Neptune or an asteroid. Even the four brightest moons of Jupiter and the phases of Venus can be detected through common binoculars such as this 10 x 50mm model.

A pair of binoculars can be thought of as two small refracting telescopes that are placed side by side, to accommodate each eye, to produce an exciting stereoscopic view. Binoculars come in a variety of sizes, the most common being 7 x 50mm and 10 x 50mm, with the latter considered the maximum size for hand-held use before shakiness becomes most apparent. In the case of binoculars rated 10 x 50mm, the '10 x' relates to the power of magnification while the '50' defines the aperture of each objective lens. Since most popular models are limited to small powers such as 7X or 10X, they are unsuitable for close-up views of the planets. However, this is not to say that they have no place in observing the Solar System.

A good quality pair of binoculars can produce stunning views of the Moon, revealing a number of larger craters and mountain ranges. Basic planetary views are also possible with steady hands or a mount. For example, 7 x 50mm binoculars will reveal Jupiter's disc and its four major satellites. The phases of Venus can also be detected, especially around inferior conjunction. Merged with its ring system, Saturn is clearly visible as a planet among the stars with an oval appearance. Binoculars are well suited to observing comets and meteor radiants such as the November Leonids. In fact, a pair of binoculars may be all you need to discover a new comet! On top of this, they are great for detecting bright asteroids that move among the background stars and the fainter planets Uranus and Neptune.

There are, of course, very expensive large (sometimes referred to as giant) binoculars with apertures up to 100mm (3.9in) or greater. These are best used with a sturdy mount for stability and enjoyable observing. The wide-angle views are simply mind-blowing.

# Useful calculations

While some beginners may not initially be concerned with formulas for measuring the optical aspects of their telescope and accessories, knowing a little about some of the most basic calculations is helpful. This is especially the case when making observational records that may be shared with others and to ensure you use only the magnifying powers your telescope is capable of in practical terms.

## Maximum practical power of a telescope

As a general rule, the maximum practical power your telescope can deliver before the image begins to degrade (a fuzzy, poor contrast appearance) is 2X magnification per millimetre of aperture or 50X magnification per inch of aperture in ideal observing conditions. Thus using the most powerful eyepiece you may have (or eyepiece and Barlow lens combination) with a telescope of, say, 250mm (10in) aperture, the maximum yielded power should not realistically exceed about 500X magnification.

## Eyepiece Magnification

So how do we know what power an eyepiece will produce? Firstly, your eyepiece will be labelled with a focal length in millimetres, for example 10mm (0.39in) or 25mm (1in). Likewise, most telescopes are labelled with a rated focal length, usually found on a label near the focusing assembly. If not, it may be noted in the user's manual. For argument's sake let's say it's 1000mm (40in) . To calculate the power produced by a 25mm eyepiece with this instrument, divide the focal length of the telescope by the eyepiece's focal length. Therefore 1000 divided by 25 will produce a magnification of 40X. In this example a distant object such as a tree on a hilltop would appear 40 times closer through the telescope than it does with the naked eye.

## f/ratio and focal length

As discussed earlier in this chapter, some telescopes are rated with an f-number, being the ratio of its focal length to the diameter of its primary mirror or objective lens. It is calculated by dividing the focal length of the primary lens (or mirror) by its diameter. Thus a 1000mm (40in) focal length telescope with a 200mm (8in) mirror has a focal ratio of f/5. If the focal length of a telescope is unknown it can be deduced simply by multiplying the known f/ratio by the diameter of the objective lens or mirror; thus 5 x 200mm equals a focal length of 1000mm.

In prime focus photography (i.e. with the camera lens removed), a camera mounted directly to this telescope would be operating at f/5. By adding a 2X Barlow lens to this configuration the focal length of the telescope is increased to 2000mm and the focal ratio will be f/10. The subject being photographed appears twice as large but within a narrower field of view.

# Filters

For planetary work, colour filters are your keys to discriminating atmospheric and surface details. Threaded eyepiece filters are readily available and generally affordable. The most common and perhaps most practical set consists of four low-cost, coloured glass filters. Standard Kodak Wratten numbers for photography rates them. They are the yellow No. 12, red No. 23A, light green No. 56 and blue No. 80A.

To give you some idea of how these useful accessories filter incoming light and enhance images at the eyepiece, the summary below provides a general guide.

## Yellow No. 12

This filter enhances features on the lunar surface. It darkens atmospheric detail containing low-hue blue tones in Jupiter's atmosphere while also enhancing orange and red features of the belts and zones. When observing Mars, it reduces light from the blue and green areas, which darken low albedo maria along with lightening the ochre hues of the surrounding desert regions. A yellow filter can also sharpen the boundaries of yellow dust clouds in regions where they may be forming. Used with Saturn, it can penetrate and darken atmospheric details that exhibit low-hue blue tones while enhancing orange and red features of the belts and zones at the same time. It can improve subtle details when observing Uranus in larger instruments of 305mm (12in) and larger.

## Red No. 23A

A very useful filter, the red No. 23A reduces light from blue and green areas, which darkens the maria and other tract markings while lightening the orange-hued desert regions of Mars. It also sharpens the boundaries of yellow dust clouds and can improve definition of comet dust tails. Red filters are especially useful for studying and imaging the bluer clouds in the atmospheres of Jupiter and Saturn. They will improve observations at twilight when observing Mercury near the horizon, and when observing Venus during daylight (reducing the background brightness of the sky). A red filter also aids detection of occasional deformations of Venus' terminator.

## Green No. 56

Green No. 56 is useful for observing the low-contrast hues of blue and red that exist in Jupiter's atmosphere and is excellent for increased contrast of Martian polar caps, low clouds and yellowish dust storms. A green No. 56 will also accentuate cloud belts in Saturn's atmosphere.

## Blue No. 80A

This filter greatly enhances the boundaries between reddish-brown belts and the adjacent bright zones of Jupiter, and is very effective for viewing the Great Red Spot. This is a very good filter for detecting high atmospheric clouds in the Martian atmosphere while also accentuating contrast of polar ice caps. It enhances low-contrast features between the belts and zones of Saturn, and is capable of detecting occasional dark cloud structures in the cloud tops of Venus (although a violet filter will offer more promising results).

## Moon filter

Moon filters are cost-effective options for reducing the brightness and glare of the Moon, particularly when it is full. They improve overall contrast so that greater detail can be observed on the lunar surface. Light transmission is about 18 per cent.

## Polarising filter sets

Polarising filter sets transmit only light moving along a specific plane, thereby increasing the contrast between surfaces with different planar transmissions. Generally, the polarisation effect makes blue skies a deeper blue, eliminates surface reflections and helps reduce atmospheric haze. Polarising filters are particularly useful for reducing glare while observing the Moon and planets, without affecting the colour of the object being observed. A set is comprised of two filters that thread into either the eyepiece or an adaptor (usually supplied with the set). When one filter is used alone it transmits about 30 per cent light. When they are used together, the degree of transmission can be varied, causing the amount of transmitted light to change. At maximum density, a polarising filter set transmits only 5 per cent of entering light.

Eyepiece filters are extremely useful for discriminating features on the surface and in the atmospheres of the planets. They can also be used to produce tri-colour images with black-and-white films and CCD cameras.

## Solar filters

Gone are the days of dangerous threaded glass solar filters for eyepieces; they have been known to crack under the intense heat from magnified sunlight and must be avoided. The safest way to directly observe the solar disc is to use filters that are placed securely over the telescope's objective. A safer alternative again is a solar projection screen. Using special filters, made from materials like Baader solar film, reduces transmitted light from the Sun by 100 000 times, transmitting only 0.001 per cent of the visible light while reflecting (essentially blocking 99.999 per cent of the unwanted light). The film also absorbs all ultraviolet rays while the coating reflects infra-red light, rendering both wavelengths absolutely harmless. The Sun appears its 'real' colour, a neutral white, and the sky adjacent to the solar limb is jet black. These filters allow you to see detail in sunspots, bright faculae near the limb and the mottled areas known as *granules*.

# Creating planetary portraits

The ability to freeze a moment of time in a photograph or video movie through either a chemical process or electronics is perhaps one of the great marvels of human ingenuity. There was once a time when a family photo or even owning a camera was limited to the wealthy and empowered. Today, we live in a world of happy snaps and cameras, ranging from expensive large format cameras to inexpensive digital technology that even comes as an accessory with a mobile phone. When using zippy little film cameras, we now think little of discarding many of the unsatisfactory pictures because it doesn't really hurt our pockets any longer. Even with modern digital cameras, one need only press the delete button and click again. What was once a rare luxury is now a mass-market consumable with little thought spared for just how amazing it is that we can capture friends and loved ones through time capsule imagery that can last throughout the ages. And of course, these very same tools can be utilised to capture wonderful celestial portraits.

Most newcomers to astronomy are often quite keen to attempt some form of astrophotography through a telescope. This might be as simple as holding an instamatic camera up to the eyepiece for a snapshot of the lunar craters or capturing a faint and distant galaxy using more sophisticated imaging tools.

The two main undertakings in astrophotography are deep-sky and planetary. The primary difference between the two is related to exposure times. Of course you can simply draw what you see in the eyepiece if preferred. This is always a good starting point since it will help improve your visual acuity by making you *really* look at the object.

Basic tools for making a celestial portrait are:

- sketching
- conventional film photography
- cooled, long exposure integrating CCD cameras
- digital camera or video using surveillance video cameras or a camcorder.

Since deep-sky targets are faint, low surface-brightness objects, they require long exposures with photographic film or cooled CCD cameras in order to build up a distinguishable image from the accumulated dim light. Thermally cooled CCD cameras are far more sensitive than off-the-shelf films, producing an image in only a fraction of the time. Depending on the target and camera type used,

M20, Trifid nebula in the constellation of Sagittarius. Through the eyepiece, such deep-sky targets appear as faint hazy patches against the sky. Photographs like this one require long exposures with fast films to reveal colour and detail. This 25-minute exposure was taken with a 250mm (10in) f/4.5 Newtonion telescope using a 35mm SLR camera and ISO 1000 colour film.

long-exposure pictures can take anywhere from a few minutes to an hour or more. They require an accurately polar aligned telescope with an ability to track a nearby guide star using a guiding scope or off-axis guider (in the case of conventional CCD imaging, this may be a CCD-based auto-guider). The Sun, Moon and planets, on the other hand, require very short exposures ranging from a fraction of a second up to a few seconds.

No matter which medium you decide to adopt, making a record of your observations is both rewarding and sometimes useful to other observers. With this in mind, your planetary portrait will be even more useful if you accompany it with a few additional bits of information, such as details of your observing location (e.g. Sydney or Perth), or better still the longitude and latitude of your site. Record the time and date the picture was taken; these are typically noted in universal time (UT), more particularly known as Greenwich Mean Time (GMT). If you prefer you can simply record your local time and, from this, others can calculate the universal time for your stated location.

It is also useful for other observers to understand a little about the conditions under which the picture was taken so they know they are comparing apples with apples. This means noting sky transparency and atmospheric or local seeing effects. You should also document details of your telescope such as type, aperture, focal length used when taking the picture, types of filters used, camera type or specifications and exposure times. This basic data will greatly assist others who may also have captured the same target near the time your picture was taken. Some people may simply be inspired by the quality of your pictures and want to know more about the equipment and technique used to produce such stunning results. Moreover, this information will be useful to you when comparing your own results. Your meticulous notes can greatly assist when assessing the combined factors that helped produced the best final picture.

## Sketches

Making a drawing requires an acute eye and a little artistic skill, but most of all an honest approach. In other words, one must be diligent in drawing only that which can actually be seen rather than trying to fill in the gaps with elaborate non-visible detail. Tools for sketching include an easel or sturdy clipboard, a pad of smooth white sketching paper, lead pencils (H or HB grade), a smudging tool (perhaps your finger) and an eraser.

When drawing a planet's disc start with a template of around 5 centimetres. In the case of Mars or Jupiter you should make careful note of any apparent phase by first drawing a thin curve around the inner edge of the template circle, then carefully shading the area with the edge tip of your pencil. Since Jupiter and Saturn rotate so rapidly any major features will be displaced quickly, so first make rough outlines of any notable features as quickly as possible then complete the more rudimentary aesthetic details later.

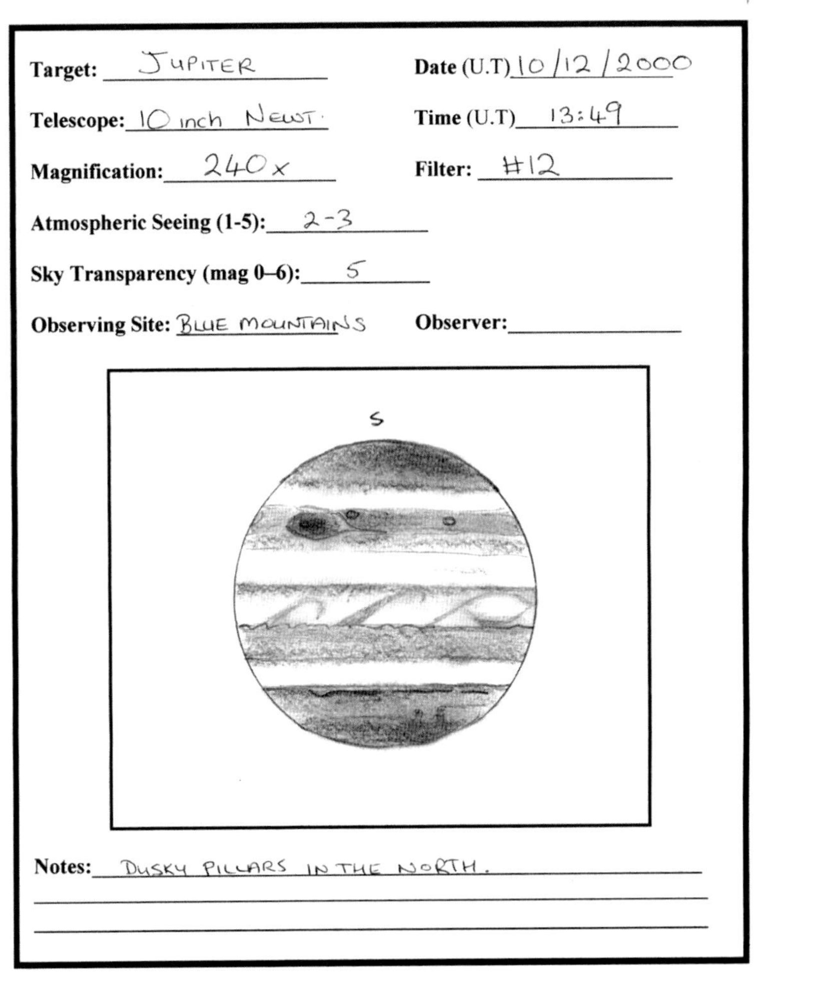

**Target:** JUPITER
**Date (U.T)** 10/12/2000

**Telescope:** 10 inch NEWT.
**Time (U.T)** 13:49

**Magnification:** 240x
**Filter:** #12

**Atmospheric Seeing (1-5):** 2-3

**Sky Transparency (mag 0–6):** 5

**Observing Site:** BLUE MOUNTAINS
**Observer:**



**Notes:** DUSKY PILLARS IN THE NORTH.

A sample observing form for sketching what is seen at the eyepiece. You don't need to be a great artist—just apply a careful and honest approach.

## Film photography

Most beginners tend to try their hand at planetary or lunar photography using a conventional film camera held up to the telescope's eyepiece—after all this seems the logical first step! This is called *afocal photography*. Although somewhat crude, this quick snapshot technique has been known to produce outstanding results particularly with bright, high-contrast subjects like the Moon.

The preferred method is called *prime focus astrophotography*. This procedure requires use of a 35mm SLR camera where the cameras lens is removed and the camera fitted with a t-ring adaptor. The entire camera can then be attached to the focuser in place of the eyepiece and the primary lens or mirror of the telescope becomes the camera's lens. Having the camera fixed to the telescope in this way is a far more stable solution than hand-held afocal photography. With the target visible in the viewfinder, the light is focused from the telescope's primary lens or mirror/s onto the camera's film plane. With short focal length telescopes, prime focus astrophotography may be okay for general pictures of the Sun and Moon, but planetary photographs will be somewhat less appealing and appear like tiny blobs with little or no detail in the final print. For planetary photography the telescope's effective focal length must be greatly increased in order to project a larger image onto the film. This can be done by attaching the camera to a 2X or 3X Barlow lens or with a specialised astrophotography eyepiece projection adaptor.

When operating a telescope at high powers, the slightest bump or strong breeze will buffet the image enough to spoil the photo. If your telescope is in an unsheltered area then choose nights when there is little or no breeze; this is especially the case when using a long tube telescope on a flimsy mount. To avoid vibrations caused when clicking the camera's shutter release button use a shutter release cable, negating the need for direct contact with the camera body itself. Even with a shutter release cable, small vibrations can still occur and it is recommended you use a black piece of cardboard held in front of the telescope for a few seconds before and while engaging the camera shutter. This will allow time for any subtle vibrations in the system to settle. Once settled, quickly remove the cardboard to start the exposure, and when completing the shot quickly place it in front again before releasing the shutter. Of course if your telescope is on a wonderfully rock-solid mount then you will not need to resort to such desperate measures. Another technique best suited to planetary photography is to utilise the camera's built-in self-timer function if available.

Wonderful wide-field photographs of the sky can be taken with a simple 35mm SLR camera mounted piggyback on a driven equatorial telescope.

A 35mm SLR camera is attached to the telescope using a camera projection adaptor. An eyepiece sits inside the adaptor barrel and projects the image onto the camera's film plane.Using the camera's self-timer shutter delay allows shakiness in the system to settle before the exposure begins.

This usually allows up to 10 seconds of delay before triggering the shutter. If the vibrations in your telescope have not settled within 10 seconds then buy a new mount!

You may be asking yourself, 'What sort of film should I use?'. This depends largely on the brightness of the object being photographed, and by this I mean more than just apparent visual brightness in magnitude. As discussed in earlier chapters, the greater the magnification used, the dimmer a subject becomes. So when an object is dim you need to allow the film more time to gather those precious photons in order to build up a nice looking picture. Alternatively, you can use films of greater sensitivity and faster response times. But there are other factors that contribute to the outcome of your celestial portrait. Since we know that atmospheric seeing can distort a picture-perfect moment within seconds, a shorter exposure time (the length of time a camera shutter is open) is also essential. But there's a trade-off: speed versus graininess. Common 35mm films are rated by an ASA or ISO speed rating. A film with a rating of 100 produces a much more aesthetically pleasing picture compared with that taken using 1000 film, the latter producing a much grainier looking picture. Since you'll want to capture as much fine planetary detail as possible, you can't afford to lose it in the graininess of faster films; thus slower films (100, 200 or 400) are best suited to the task.

Well known to the veteran deep-sky astrophotographer is the benefit of keeping film cold. The chemical coatings on photographic film respond more efficiently when freshly loaded for immediate use after being stored in the fridge or freezer. This is one technique you may want to consider in order to reduce exposure times and preserve the longevity of unused film.

Even a seasoned photographer can expect to waste many rolls of film before achieving a satisfactory result. Conventional photography of the planets is notoriously difficult and subject to the random warping effects of poor seeing. Good film photography of the planets requires an accurately polar aligned, medium- to large-aperture telescope, with a sturdy driven mount and excellent vibration dampening characteristics. Just as importantly, it requires excellent seeing, patience, practice and perseverance. Since most entry-level photography is a process of trial and error, always bracket your exposures to ensure that at least one picture will prove worthwhile.

The table opposite is a guide to recommended exposures, in seconds, for the planets and the Moon.

**Moon—thin crescent** | *Focal ratio*

| ISO | f/2.8 | f/4 | f/5.6 | f/8 | f/11 | f/16 | f/22 | f/32 | f/64 |
|---|---|---|---|---|---|---|---|---|---|
| 100 | 1/125 | 1/60 | 1/30 | 1/15 | 1/8 | 1/4 | 1/2 | 1sec | 2sec |
| 200 | 1/250 | 1/125 | 1/60 | 1/30 | 1/15 | 1/8 | 1/4 | 1/2 | 1sec |
| 400 | 1/500 | 1/250 | 1/125 | 1/60 | 1/30 | 1/15 | 1/8 | 1/4 | 1/2 |

**Moon—quarter**

| ISO | f/2.8 | f/4 | f/5.6 | f/8 | f/11 | f/16 | f/22 | f/32 | f/64 |
|---|---|---|---|---|---|---|---|---|---|
| 100 | 1/500 | 1/250 | 1/125 | 1/60 | 1/30 | 1/15 | 1/8 | 1/4 | 1/2 |
| 200 | 1/1000 | 1/500 | 1/250 | 1/125 | 1/60 | 1/30 | 1/15 | 1/8 | 1/4 |
| 400 | 1/2000 | 1/1000 | 1/500 | 1/250 | 1/125 | 1/60 | 1/30 | 1/15 | 1/8 |

**Moon—full**

| ISO | f/2.8 | f/4 | f/5.6 | f/8 | f/11 | f/16 | f/22 | f/32 | f/64 |
|---|---|---|---|---|---|---|---|---|---|
| 100 | 1/2000 | 1/1000 | 1/500 | 1/250 | 1/125 | 1/60 | 1/30 | 1/15 | 1/8 |
| 200 | >1/2000 | 1/2000 | 1/1000 | 1/500 | 1/250 | 1/125 | 1/60 | 1/30 | 1/15 |
| 400 | >1/2000 | >1/2000 | 1/2000 | 1/1000 | 1/500 | 1/250 | 1/125 | 1/60 | 1/30 |

**Sun—through solar filter**

| ISO | f/2.8 | f/4 | f/5.6 | f/8 | f/11 | f/16 | f/22 | f/32 | f/64 |
|---|---|---|---|---|---|---|---|---|---|
| 100 | 1/1000 | 1/500 | 1/250 | 1/125 | 1/60 | 1/30 | 1/15 | 1/8 | 1/4 |
| 200 | 1/2000 | 1/1000 | 1/500 | 1/250 | 1/125 | 1/60 | 1/30 | 1/15 | 1/8 |
| 400 | >1/2000 | 1/2000 | 1/1000 | 1/500 | 1/250 | 1/125 | 1/60 | 1/30 | 1/15 |

**Mercury**

| ISO | f/2.8 | f/4 | f/5.6 | f/8 | f/11 | f/16 | f/22 | f/32 | f/64 |
|---|---|---|---|---|---|---|---|---|---|
| 100 | 1/1000 | 1/500 | 1/250 | 1/125 | 1/60 | 1/30 | 1/15 | 1/8 | 1/4 |
| 200 | 1/2000 | 1/1000 | 1/500 | 1/250 | 1/125 | 1/60 | 1/30 | 1/15 | 1/8 |
| 400 | >1/2000 | 1/2000 | 1/1000 | 1/500 | 1/250 | 1/125 | 1/60 | 1/30 | 1/15 |

**Venus**

| ISO | f/2.8 | f/4 | f/5.6 | f/8 | f/11 | f/16 | f/22 | f/32 | f/64 |
|---|---|---|---|---|---|---|---|---|---|
| 100 | >1/2000 | 1/2000 | 1/1000 | 1/500 | 1/250 | 1/125 | 1/60 | 1/30 | 1/15 |
| 200 | >1/2000 | >1/2000 | 1/2000 | 1/1000 | 1/500 | 1/250 | 1/125 | 1/60 | 1/30 |
| 400 | >1/2000 | >1/2000 | >1/2000 | 1/2000 | 1/1000 | 1/500 | 1/250 | 1/125 | 1/60 |

**Mars**

| ISO | f/2.8 | f/4 | f/5.6 | f/8 | f/11 | f/16 | f/22 | f/32 | f/64 |
|---|---|---|---|---|---|---|---|---|---|
| 100 | 1/1000 | 1/500 | 1/250 | 1/125 | 1/60 | 1/30 | 1/15 | 1/8 | 1/4 |
| 200 | 1/2000 | 1/1000 | 1/500 | 1/250 | 1/125 | 1/60 | 1/30 | 1/15 | 1/8 |
| 400 | >1/2000 | 1/2000 | 1/1000 | 1/500 | 1/250 | 1/125 | 1/60 | 1/30 | 1/15 |

**Jupiter**

| ISO | f/2.8 | f/4 | f/5.6 | f/8 | f/11 | f/16 | f/22 | f/32 | f/64 |
|---|---|---|---|---|---|---|---|---|---|
| 100 | 1/500 | 1/250 | 1/125 | 1/60 | 1/30 | 1/15 | 1/8 | 1/4 | 1/2 |
| 200 | 1/1000 | 1/500 | 1/250 | 1/125 | 1/60 | 1/30 | 1/15 | 1/8 | 1/4 |
| 400 | 1/2000 | 1/1000 | 1/500 | 1/250 | 1/125 | 1/60 | 1/30 | 1/15 | 1/8 |

**Saturn**

| ISO | f/2.8 | f/4 | f/5.6 | f/8 | f/11 | f/16 | f/22 | f/32 | f/64 |
|---|---|---|---|---|---|---|---|---|---|
| 100 | 1/125 | 1/60 | 1/30 | 1/15 | 1/8 | 1/4 | 1/2 | 1sec | 2sec |
| 200 | 1/250 | 1/125 | 1/60 | 1/30 | 1/15 | 1/8 | 1/4 | 1/2 | 1sec |
| 400 | 1/500 | 1/250 | 1/125 | 1/60 | 1/30 | 1/15 | 1/8 | 1/4 | 1/2 |

# CCD imaging

Developed in 1970, a CCD (charged coupled device) is the image sensor found in all modern camcorders and digital cameras. Conventional films record light by chemically reactive processes whereas a CCD converts photons into an electric charge. A CCD chip houses an array of thousands or millions of light-sensitive cells called pixels. In very simple terms, each pixel registers and converts this light into a numeric value of intensity which is then converted through digital processing to create a picture. The output of a CCD camera may interface directly with a computer via a serial or USB port connection. Low-cost CCD video cameras produce a composite output signal that can be interfaced to a computer via a video capture card or port or directly to a video cassette recorder.

The imaging chip at the heart of a CCD camera. When the camera is not in use, always keep a cap fitted over the opening of the camera to avoid tiny dust paritcles settling on the CCD. Dust specks can appear the size of boulders on the television monitor.

Astronomical (still image) CCD cameras are thermally cooled, integrating devices, with a greater dynamic range than uncooled cameras. They are generally more expensive but well worth the investment should you decide to pursue serious astrophotography, especially deep-sky work. While operational simplicity and set-up has improved greatly in these cameras over the years, if you're a beginner you'll probably find more instant satisfaction from less expensive and more readily available video and digital still cameras; in keeping with the introductory nature of this book we'll discuss the use of these more economical tools and their specific advantages in lunar and planetary imaging. If you wish to find out more about advanced astronomical CCD cameras, contact your local astronomy dealer or consult the wealth of information available over the Internet.

# Digital cameras

Modern CCD image sensors are now more efficient at recording photons from dim light sources than ever before and video or digital still cameras are by far the simplest and most cost-effective means of recording bright Solar System targets. No waiting for films to be processed; you can see the results right before your eyes.

A digital still camera is a convenient alternative to a 'wind and click' camera, and there are adaptors specifically designed to attach a digital camera to the eyepiece of your telescope. The main drawback for astrophotography with most economical models is the inability to remove the camera lens. This limits the user to afocal photography and thus a little bit of valuable light is lost through the interference of an additional lens in the light path. The same limitation exists in most modern camcorders. Despite this slight drawback though, an 'optically' zoomed digital camera or camcorder aimed at the eyepiece of a telescope quickly reveals the benefits of larger image scale and instant results are achievable. Note that 'optical zoom' is not digital zoom! What's the difference you ask? Optical zoom

Left: the Kodak WebCam is stripped down and converted into a fully fledged astrocamera, pictured at centre. Most ball-shaped webcams today can be easily utilised in their existing enclosures by simply removing the lens and adding a 31.8mm (1.25in) piece of tube so it can be inserted into the focuser of a telescope. Right: a simple hand-held digital camera can be held up to the eyepiece for quick snapshots of the Moon and planets.

is true magnification of light via a series of lenses for closer inspection, while digital zoom artificially enlarges the picture by pixel interpolation and cropping, giving the effect of a closer view. The truth of the matter is that no additional detail is produced with digital zooming and when used beyond a certain point a picture can take on a less-than-appealing look.

Most economical cameras also have automatic exposure control, making it difficult to obtain the optimal exposure for a tiny planet set against a dark background. A camera that will allow you to adjust the exposure manually is preferable.

Aside from the conventional digital still camera we are all familiar with, there is yet another, even more economical and convenient entry-level imaging tool. The desktop PC camera or webcam has been around for several years now and is being widely adopted by amateurs around the world to produce wonderful pictures of the Solar System and in some cases the brighter deep-sky targets too. With a little carefully handled modification, these cameras can be mounted directly to the focuser of a telescope and simply connected to the USB port of a laptop or desktop computer. Popular models include the Kodak desktop cameras, and the Philips ToUcam and QuickCam among many other brands. These cameras are capable of taking single snapshots and even movies. The internal lens can be removed to expose the tiny CCD chip inside, then all you need to do is glue a short length of 31.7mm (1.25in) PVC or aluminium tubing to the body and there you have it—instant astrocam!

## Video astronomy

Real-time video has been a major boon for amateur lunar and planetary imaging in the last 10 years. Not only is it affordable, but the technical advantages for obtaining a single, almost perfectly undistorted view of the planets and lunar craters are many.

The four major benefits of video are: instant results; large image scales over conventional 35mm film; thousands of individual short exposures in minutes; and individual picture selection. One hour of videotape can record over 90 000 individual pictures for less than the cost of purchasing and processing a roll of 24-exposure photographic film. Furthermore, the fast frame rates of video means we can freeze the seeing by increasing the odds of capturing one or more images relatively undistorted by atmospheric turbulence. PAL video can snap off twenty-five images to tape in just one second and if you're unhappy with the results you can simply erase it and start again.

This picture depicts how objects appear on a television screen or monitor using video cameras with different-sized image sensors. Mounted to a telescope operating at f/25, from left to right the CCD chip sizes are ½in, ⅓in and ¼in respectively. For planetary work, telescopes with shorter focal lengths will benefit from cameras with smaller image sensors, thus producing larger image scales.

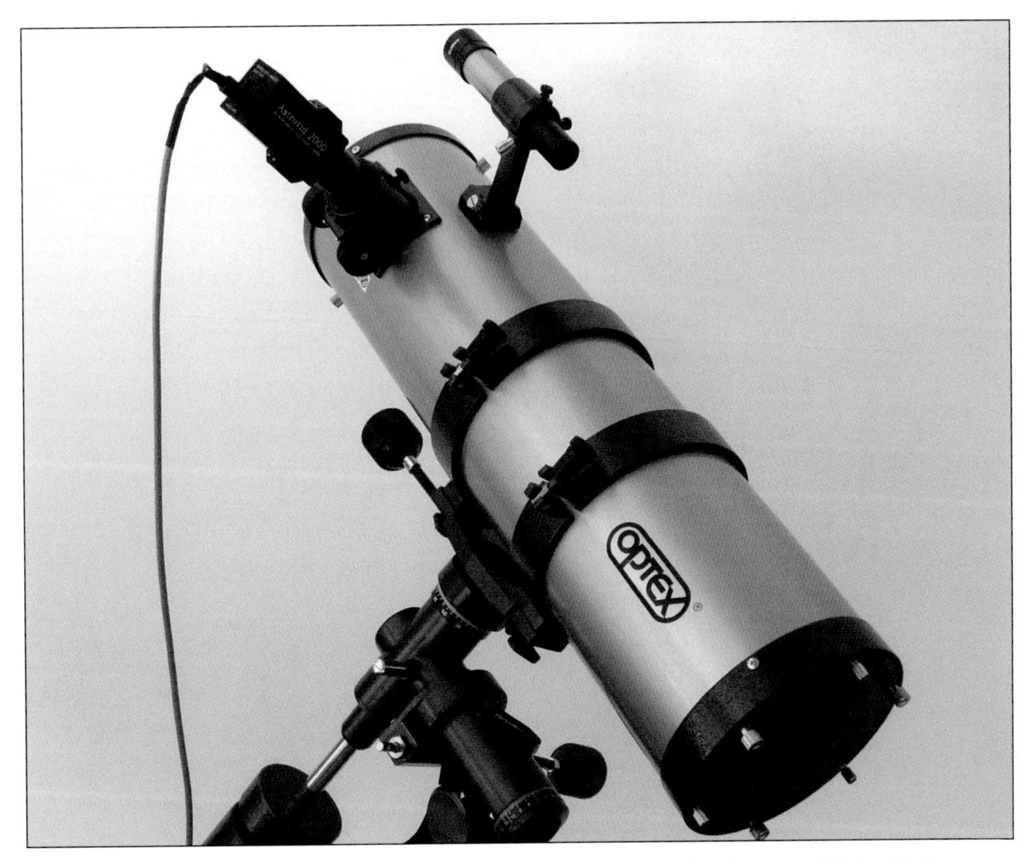


The Astrovid 2000 camera is mounted directly into the focuser of a telescope. With a 31.7mm (1.25in) adaptor supplied, the camera can also be inserted like an eyepiece into a Barlow lens for greater magnification.

## Surveillance video cameras

These low-cost security cameras, also known as closed-circuit television (CCTV) cameras, and camera modules can be purchased through most electronics dealers for as little as $40 to $70. They have easily removable threaded lenses and allow you to view the planets on your television screen. A tiny, lightweight CCTV camera coupled to a Barlow lens or eyepiece projection adaptor can reveal large swirling festoons, fine wispy details and dark knots within belts of Jupiter that will hold your attention for hours. Watching sunspots or the lunar landscape in real-time on your television screen is the best eye relief you could wish for. Video is also a wonderful medium for public or club displays on a large projector screen or for simply sharing with the family at home on cold winter nights.

The best type of video camera for lunar and planetary imaging is one that enables you to manually adjust both the shutter speed rate and signal gain. Most off-the-shelf black-and-white surveillance cameras these days are pre-programmed for automatic adjustment of these functions and this can sometimes be annoying, especially for planetary work. The ability to manually adjust these settings offers greater control of recorded image contrast and depth, brightness and background noise levels in the picture.

The number of television lines a video camera outputs governs the resolution of the picture it produces. The higher the number, the finer the detail displayed; most are typically rated between 400 to 600-plus lines. The sensitivity of a CCD image sensor is measured in terms of a lux value for a given f/ratio. The

The Astrovid range of cameras from Adirondack in the United States have been very popular among amateur astronomers. They are simply a conventional low-light security camera modified to allow manual control of shutter speed and signal gain for optimal exposures.

smaller the number of its lux rating, the more sensitive is the CCD chip, which is better for our purposes. The ideal camera for astronomical use is one that produces lots of television lines, has a low lux rating of 0.1 or better, and manual shutter speed and signal gain controls. Typical black-and-white output surveillance video cameras range from 0.1 to 0.05 lux at f/1.2. For lunar and planetary imaging, compact video cameras are now so inexpensive that there is little excuse for pursuing the more difficult task of conventional film photography, especially since they produce far superior results, with greater convenience and flexibility.

Just like the prime focus photography method discussed earlier, you can adapt a video camera in exactly the same way, only the incoming light is focused onto a CCD chip at the heart of the camera. A single Barlow lens or stacked combination can be used to magnify the image projected on the CCD for a more dramatic result. For the planets, best results are achieved with focal ratios of around f/20 to f/30 and greater, depending on the sensitivity of your camera.

## Using a camcorder

Before splashing out to purchase a low-light security camera or disassembling your webcam, you may already have a convenient mini television studio in the form of a camcorder tucked away in the cupboard. No messy cables to run, just point the camcorder into the eyepiece and press the record button! This is a great way to start out and you'll be astounded at the stunning views of the Moon or sunspots you'll be able to record. Another benefit is the ability to zoom in on features with leisurely convenience.

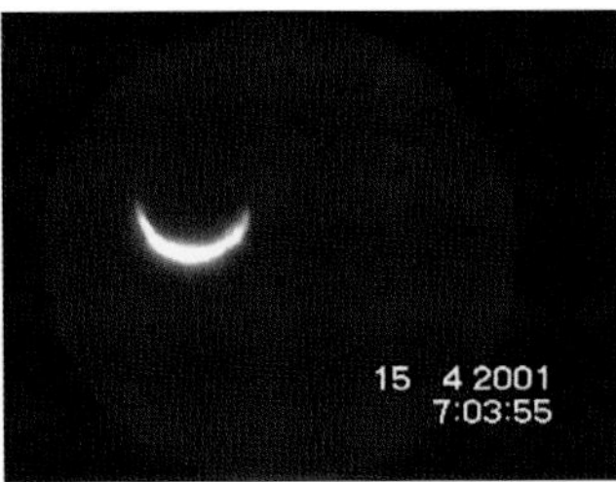

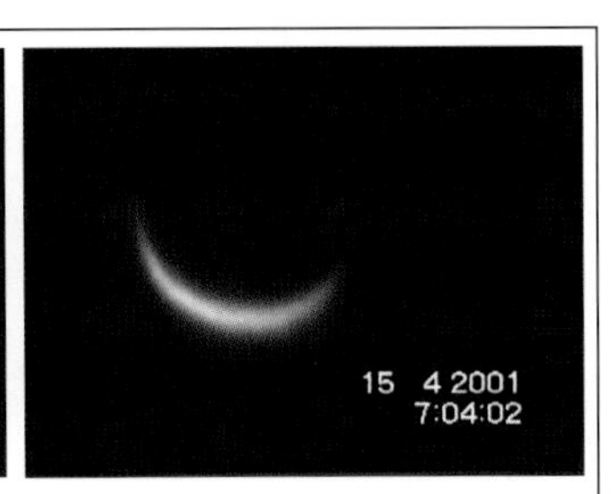


This series of images demonstrates the effects seen when placing a video camera up to the eyepiece of a basic 60 x 700mm refracting telescope to capture the planet Venus. As the camera's zoom function is engaged, evidence of the outer edges of the eyepiece and vignetting caused by the internal field stop becomes less apparent.

If your telescope isn't equipped with a drive motor then examining a planet can be somewhat frustrating since the diurnal motion of the Earth is so much more apparent at higher magnifications. You will often find, when fumbling around to locate a planet through a very narrow field of view, that it has drifted from the eyepiece. Moreover, the novelty of holding a camcorder, with steadiness, up to the eyepiece can quickly wear thin. Just like digital still cameras, adaptors and special brackets are also available for attaching a camcorder to a telescope.

The light sensitivity of modern-day camcorders has certainly jumped ahead in leaps and bounds since the 1990s, and digital cameras have superior resolution compared to the old Video 8 and Hi-8 models. In terms of performance in low-light situations, you may encounter camcorders offering a '0-luxnight shot' feature or similar. By engaging this function, an infra-red blocking filter within the camera is removed from the light path entering the camera, thus increasing its response to normally invisible near infra-red light. At the same time, an infra-red LED (light emitting diode) on the exterior housing of the camera illuminates a small area in front of you, which is now detectable by the imaging chip. Since this feature is designed for dark external situations, it has no 'real' useful effect when imaging subjects through a telescope.

An important tip to remember is to switch off the auto-focus mode of your camera. If left on you will find it jumping in and out of focus with frustrating persistence. Manually adjust your camera to the infinity focus setting, usually symbolised by an icon of a mountain peak or similar.

## Recording video images

The benefit of direct video capture to a computer or digital tape is the ability to record all the precious resolution that the camera is capable of producing. In some models this may be up to 800 lines. Conventional VHS machines record a limited amount of this information while the more superior SVHS systems can record around 400 television lines. Since computers have greater processing power and larger hard drives these days, fewer video frames are dropped during the capture process due to memory restrictions and disk space is less of an issue, but a disk drive can still fill up quickly. If using a computer it is best to limit capture sequences from 3- to 5-second bursts, especially when capturing full resolution video (768 x 576 pixels) uncompressed. This yields seventy-five to 125 individual 'stills' and by employing YUV-422 compression (a common option supplied with software provided with video capture cards) or similar, file

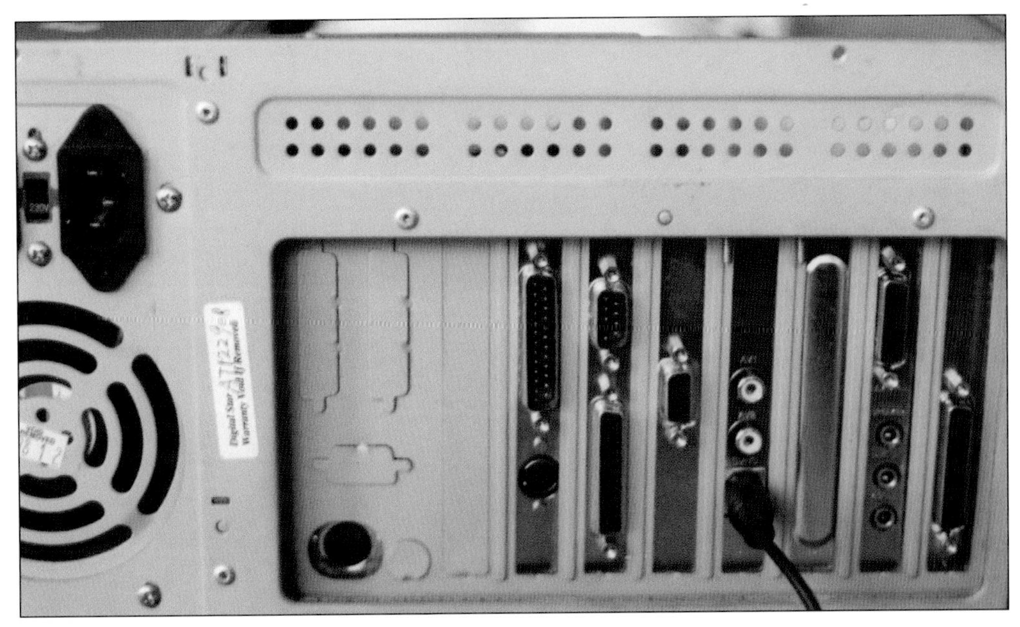

Video capture cards are very affordable today and can be placed in a spare computer slot with little difficulty. The rear of the card seen here offers two composite and one S-video connection.

sizes can be reduced significantly. If the disk begins to fill quickly then files can be easily transferred to CD-ROM to free up valuable space.

With the appropriately installed software, the USB connection in many computers is an alternative port for capturing video direct from your camera or VCR, using an appropriate adapter or a desktop camera modified for use with a telescope as previously discussed.

## Making colour images with a black-and-white video camera

A picture tells a thousand words, so the expression goes, and there is something to be said for the aesthetically pleasing look of a colour picture of Mars, Jupiter or Saturn over one which is monochrome (black and white) only. This is not to say that a black-and-white image has no aesthetic appeal. The truth of the matter is that both formats produce a unique and visually stimulating perspective while also providing useful or interesting information about a planetary subject.

To an experienced astrophotographer, the term 'tri-colour imaging' is no mystery. The technique has been commonly employed for decades to produce colour

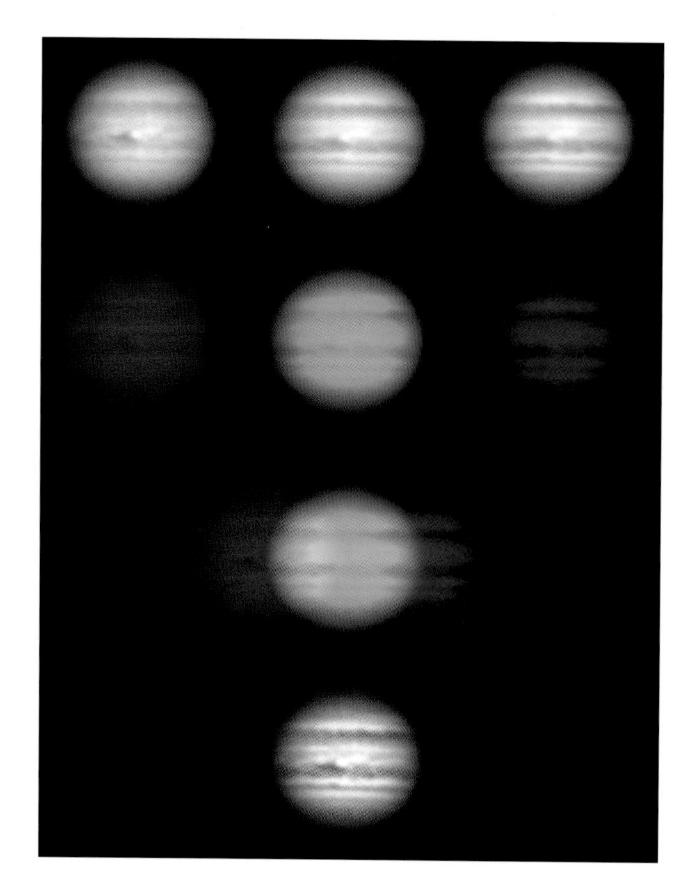

Here you can see how a colour image is produced from black-and-white video camera images taken through coloured filters. At top are the monochrome results from respective red, green and blue filters. Below, each 8-bit monochrome image is combined with its associated colour channel using image-processing software such as Adobe Photoshop to produce the final 24-bit image at bottom.

pictures from black-and-white films and image sensors. However, when it comes to a black-and-white video camera, I have often been asked, 'How did you get a colour picture of Jupiter from a black-and-white video camera?' The answer is simple, but requires the use of image-processing software to create the final colour picture.

To create a basic colour picture, the three primary colours (red, green and blue—RGB) are combined to produce the various hues for each colour of the visible spectrum, like the colours we see in a rainbow. By capturing successive black-and-white images using appropriate red, green and blue light filters, each resulting image can then be assigned to its respective colour channel and combined to create the final colour picture using image-processing software. This also preserves valuable spatial resolution.

Inexpensive colour cameras often produce poorer image contrast and increased signal noise under low-light conditions when compared to the results of a monochrome output camera. In such cameras a colour CCD chip is simply a black-and-white image sensor with tiny red, green and blue strips placed along each row of pixels, and spatial resolution is sacrificed. The preferred option is a colour video camera that contains three CCD chips. These cameras are substantially more expensive but do preserve valuable spatial resolution. As a final note, most black-and-white security-style cameras are quite sensitive to the near infra-red part of the spectrum. While it is not mandatory, an infra-red blocking filter can be used in conjunction with the coloured filters to produce more accurate colour results.

## Time lapse animations

Watching the progress of something that usually unfolds over several hours or even days (for example the blooming of a rose bud) within a few seconds is always a stunning and educational experience. Most of our brighter celestial neighbours are alive with activity and for a time lapse project there are plenty of subjects to choose from. They might include the changing features across the globe of a planet as it rotates over several hours or the dance of the Galilean moons, including transit events mentioned in earlier chapters. You might want to make a movie of sunspots as they traverse the face of the solar disc, or the changing shadow peaks of mountains across the vast lunar plains. You can even track the progress of a bright asteroid as it moves among the stars.

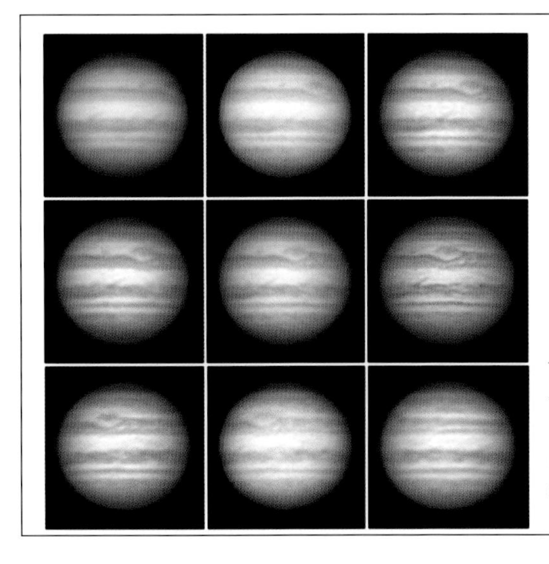

This sequence of Jupiter shows the passage of the Great Red Spot (from top left to bottom right) across the disc as the planet rotates. Using movie-making software, this selection of images makes a fine animated movie.

The key to creating attractive time lapse animations is to first capture several evolving changes in a subject at evenly timed intervals. This is important for producing smooth running transitions in the final movie or video. Timing can be critical, particularly if doing tri-colour imaging in the case of a planet such as Jupiter, which rotates quickly. Always try to keep the subject well centred in the frame during each exposure or capture session, thus making alignment easier in the final animation.

## Planning ahead

Good planning will ensure your observing and recording session is enjoyable and fruitful. Most serious amateurs plan a session weeks, months or even years ahead and planning for an upcoming event usually starts with an annual sky guide or computer planetarium program. Periodicals such as club newsletters or astronomy magazines are also useful. In addition, there is a wealth of Internet sites offering up-to-the-minute news and information.

All you need to do is decide well in advance which event you want to record and write it down on a calendar. Mark the start and end times then make a note as to how many hours in advance you may need to start setting up your equipment. Write a checklist of items you will need, especially if you're transporting equipment to a remote location. Things such as writing material, power supplies, cables and film are often forgotten.

If you are using a video recorder outdoors be sure to protect it from any dew that might condense on the delicate internal mechanisms. The absence of such protection may result in poor recording and playback quality along with possible damage to internal circuits.

## Final checks

If you don't have a permanent observatory, be it a fixed pier, a shed, or a dome, then you will need to ensure your telescope is correctly polar aligned and tracking nicely each time you set up. Check all connections to the computer and/or camera recording equipment for correct operation, and allow ample time for your telescope to acclimatise with the outside air temperature. At this point, you'll want to have at least an hour before your intended deadline time. This is your opportunity to detect and iron out any possible problems.

If doing conventional photography, check that the film is loaded correctly and is being wound on after a blank trial photo. Believe me, the discovery that your blood, sweat and tears pictures never saw the light of one heavenly object can be quite distressing.

If using a small security-type video camera and separate VCR you might need to check for any induced electrical interference visible on the monitor. If so, the cause may be the telescope's motor drive or other nearby electrical appliances. If this is the case first try moving cables around and clear of power cords in order to reduce the amount of visible noise. Moreover, turn off any non-critical electrical appliances if a noisy image persists. Check that cables are correctly terminated and that there is no corrosion on the surfaces of connectors. If using a digital camera you should make sure that there is plenty of memory available or a spare memory stick to record all the pictures you intend to take.

## Framing and exposure

Carefully consider how you will frame your target and orient the camera angle accordingly. For example, you might like the rings of Saturn or the equatorial belts of Jupiter displayed horizontally or at a particular angle. If you plan to make time lapse animations then consistency of orientation will make later image-processing much easier. Once satisfied with the camera angle, check the brightness and contrast or exposure setting of your camera. When using video, it is especially important to check the brightness and contrast settings of the video monitor to ensure that the recorded signal to videotape will not be over- or underexposed.

In the case of an occultation event, such as Saturn with the Moon, two different settings on the video camera will be required to maximise planetary detail. These settings depend on which edge of the Moon the planet may be entering or leaving. To achieve a stunning picture the most important factor is correct exposure, so that Saturn and its rings will not appear as an overexposed blob in the final picture. If ingress or egress occurs at the bright lunar limb, an automatic camera may start battling for the best shutter speed and gain setting in an attempt to achieve the best compromise of black background versus the bright limb. The most exciting results are when well-defined features are present in both objects. With a bit of patience and experimentation this is achievable. A manually controllable camera will make this task a breeze. If the planet is approaching or leaving the night side of the Moon then optimum brightness and contrast settings need only be made for the planet.

## Bringing your pictures to life: processing images

After capturing your video stills or pictures from a digital camera to a computer, processing these images with a view to improvement is the next option. Image processing can be highly rewarding, especially if you have good raw images to work with. Careful processing can bring to life an otherwise bland looking picture. Plus, there is often much more information contained within a raw image than meets the eye. At the same time it is important not to over-process a picture by giving it an artificial look or by introducing unwanted 'artefacts'. Revealing the otherwise hidden detail requires software tools that can enhance the contrast or sharpness of a digitised picture. But before we can do this it is important to start with good quality images. There is little to be gained from working with sub-quality raw images and trying to improve on them.

By sifting through each of the pictures stored in a digital camera, on disk or on videotape, the first thing to do is select and save only those individual images least affected by atmospheric turbulence. Once each good image is selected they can then be saved to your computer's hard disk or a CD-ROM. Selecting the best individual stills from videotape requires tape playback in step by step or frame by frame 'paused' mode. Once a clear image becomes apparent you can then capture and save that image to disk using a video capture card and software.

## Making a good picture even better

A single digital camera snapshot or video frame captured to your PC can often be good enough to leave as is (no further enhancement required) particularly in the case of large, bright subjects like the Moon. But smaller and dimmer targets like a planet can have a speckled appearance. For example the pixels in the black areas of the surrounding sky are not all black—some may appear white or somewhere in between with an overall effect much like that of a grainy photograph. This digital graininess is due to thermal noise generated within the electronics of an uncooled camera (known as background noise). Specialised cooled astronomical cameras produce much less noise in a picture than do low-cost economical digital still cameras and video cameras. In the case of video this noise is particularly noticeable when videotape playback is paused. There are other inherent factors that contribute to the apparent noise but they fall outside the scope of this book.

## Reducing the noise

A commonly practised technique for producing a smoother looking picture is known as *image stacking*. This involves using several of the original (noisy) single images to create one new image with the overall result being a smoother (less noisy) picture. With the quality of the new image being a vast improvement on the originals used to make it, use of further picture enhancement tools such as contrast, brightness and sharpening are much more effective in producing a more aesthetically pleasing final result.

Software such as Adobe Photoshop or Paint Shop Pro, among many others, have tools allowing the user to create a layered image (one which can contain all those noisy images) that can then be saved as the improved new picture. You can refer to your software help file for more information about using this tool. A much simpler technique is to download a free copy of user-friendly software like Registax or Astrostack from the Internet. These programs automate the process of gathering all your pictures from stored single-image files or a video file along with the aligning and stacking routines.

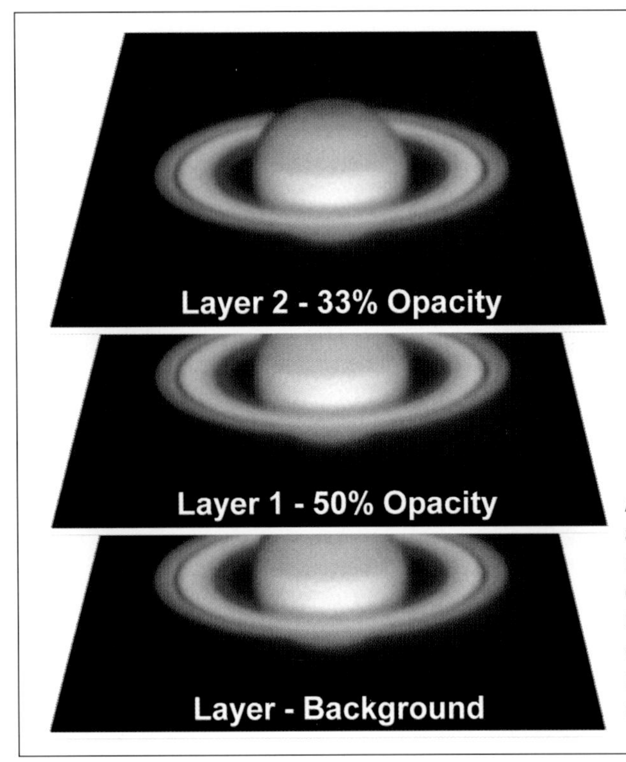


A single frame of video often appears speckly due to thermal noise within the video camera's electronics. By stacking several images as layers of different opacity, much of this background noise can be reduced, producing a much smoother final result.

The basic rule in creating the smoothest looking image is to have as many closely matched quality originals for stacking—the more the better.

Finally, always save your images in uncompressed formats such as tiff or bmp in order to preserve quality. Saving and resaving in compressed formats such as jpeg eventually reduces image quality and integrity so that they cannot be worked on reliably at some later point.

## Enhancing image detail

All image-processing software packages contain a variety of tools for manipulating pictures. Most of these functions are geared towards creative artistic use, however for processing pictures of the Sun, Moon or planets we need only concern

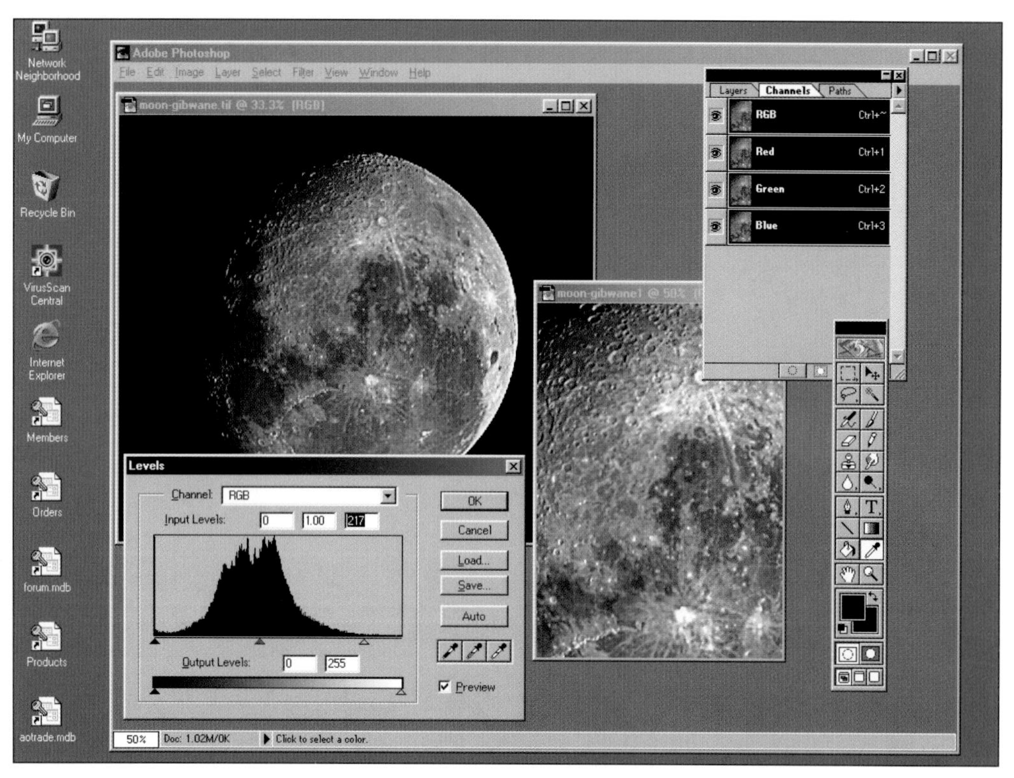


Adobe Photoshop is a terrific program for processing your celestial portraits. The x axis of the levels in the histogram seen here represent the values from darkest (0), like black sky, at the far left to brightest (255) far right, while the y axis represents the total number of pixels with that value. Under the histogram three sliders allow for level adjustment.

ourselves with the more rudimentary functions. These include the basic brightness and contrast functions and colour balancing for colour images. But brightness and contrast adjustments alone won't always bring out the subtle detail without affecting the image in some other unwanted way. So most programs provide a darkness mid-tone highlight adjustment tool for greater flexibility.

Perhaps the most important adjustments you will make are changes to the brightness values of the middle range grey tones, and this can be done without dramatically altering the shadows and highlights. These middle grey areas are where nearly all of the fine wispy detail lives in planetary images. Adjustments to these values may be all you require to reveal the natural detail present in the

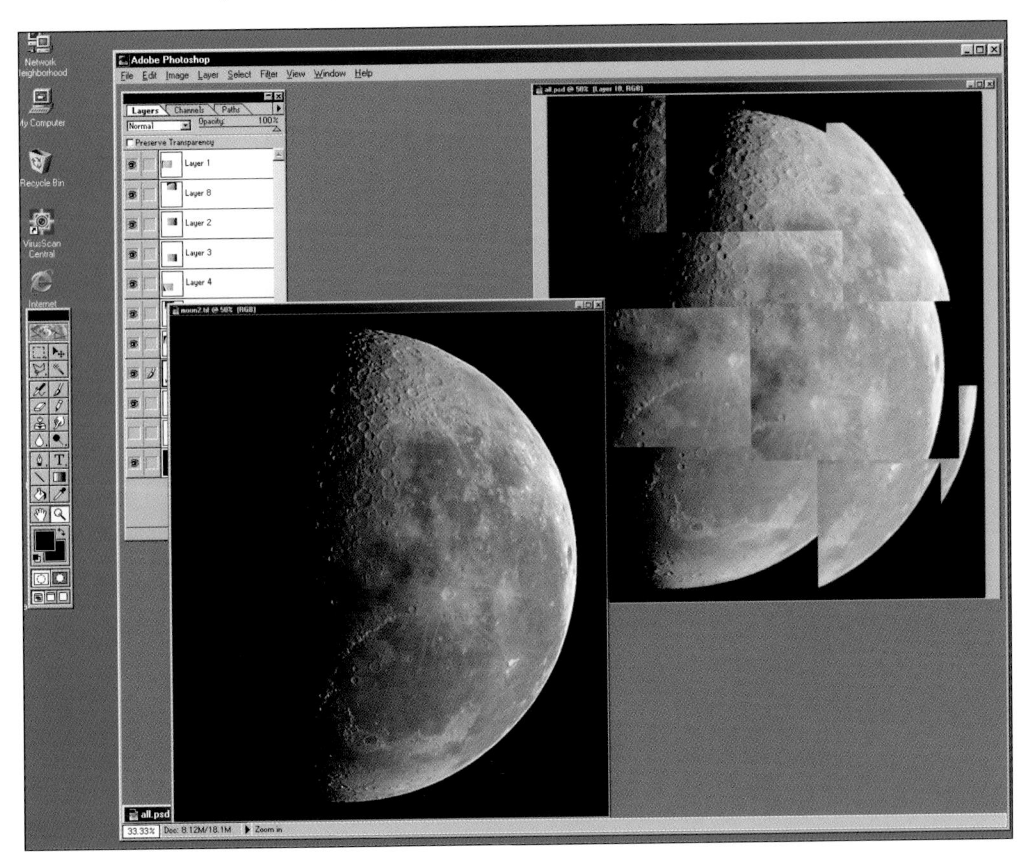


To create a high-resolution image of the Moon, several frames can be combined like a jigsaw, using the layer tool of your image-processing software. To avoid obvious boundary lines appearing in the final picture, adjust the brightness levels accordingly for closest match with the overlapping frame.

image, but there are other supplementary functions you can also explore. Perhaps the most useful filter is the 'unsharp mask' which is used to improve images that are slightly blurred. Since the atmosphere has this effect on nearly every image we take, this tool can provide excellent results but must be applied with caution. Over-use can greatly highlight any background noise, resulting in artefacts being introduced into the picture that have no place in reality.

Aside from the plethora of common image-processing programs on the market today, there are several specialised packages specifically designed for improving and analysing astronomical images, whether derived from a cooled CCD camera, digital still or video camera or even a scanned photograph. Some of these are also available freely over the Internet.

No matter which method of recording the sky you choose or how you plan to process the results, you will have captured a unique moment in time. Our celestial neighbours are undergoing constant changes, ranging from aspects of Sun angles and polar tilting to sunspots, atmospheric changes and weather pattern deviations. There will be times when perhaps you are the only one taking a picture of a particular subject at that specific moment, thus making your image unique. Always review your results with a view to improving your technique and skills, or simply to share them with others. Your pictures may even be good enough for publication in one of the many astronomy magazines published around the world. Many amateurs display their images on personal homepage websites and it is here that someone may review and request to publish your work. For quick browsing and web page loading, be sure to save images you intend to display on the Net using the jpeg compression format.

Give it a try and be sure to record all the important information discussed previously. Most of all, delight in your personal contact with the night sky and the family of worlds that occupies our neighborhood—the Solar System.

# Glossary

**Aberration:** Defect in the image formed by a lens, mirror or optical system. Spherical aberration results when different rays of light are brought to more than one focus, producing a blurred image or coma; chromatic aberration is when different wavelengths within a ray of light are brought to more than one focus, producing an image distorted by coloured fringes.

**Accretion:** A process whereby small colliding particles in space coalesce to form larger bodies through gravitational attraction.

**Airy disc:** The central spot in the diffraction pattern of the image of a star at the focus of a telescope.

**Albedo:** The ratio of the amount of light reflected by the entire lit side of an opaque body to the amount of light actually falling on it.

**Angular size:** The angle subtended by an object in the sky, such as the Moon, which has an angular size of around 30 arc minutes.

**Aphelion:** The point in a planet's orbit when at its greatest distance from the Sun.

**Apogee:** The point at which a body in orbit around the Earth, such as the Moon, reaches its farthest distance from the Earth.

**Apparent magnitude:** Measure of the observed brightness of a celestial object as seen from the Earth.

**Arc second:** A unit of angular measure of the sky. Defines the apparent visual size of a planet or separation of an object with respect to another object or reference point. An arc second is symbolised by a double quote mark (″) and an arc minute (made up up of 60 arc seconds) by a single quote mark (′).

**Asteroid:** A small rocky body orbiting a star such as our Sun. In our Solar System most of the asteroids are found between the orbits of Mars and Jupiter.

**Azimuth:** The measurement of an object's position in the sky, expressed in degrees eastward from 0° north through 360° in a horizontal plane.

**Bow shock effect:** Particles of the solar wind deflected around a planet due to the planet's invisible magnetic field.

**Conjunction:** The alignment of two objects, as seen from Earth, that appear in almost the same position in the sky. The inner planets are at inferior conjunction when situated between the Earth and the Sun. They are said to be at superior conjunction when on the opposite side of the Sun to Earth. The outer planets are said to be at conjunction when on the opposite side of the Sun from Earth.

**Cusps:** The tips or points of a crescent Moon or inferior planet.

**Declination (dec):** The term given the angular distances in degrees from the celestial equator (0°) to the celestial north pole (+90°) or celestial south pole (–90°).

**Dichotomy:** The point in time that an inner planet or our Moon appears exactly half lit.

**Eccentricity:** The amount by which an object's orbit deviates from a perfect circle.

**Eclipse:** When one body passes in front of another, such as the Moon with the Sun. Eclipses can be total or partial.

**Ecliptic:** An imaginary line across our sky that the Sun travels along on its annual passage against the background stars. It is in fact the plane of the Earth's orbit around the Sun projected on the sky.

**Egress:** The end of a transit event.

**Elongation:** A measure of the angular separation between the Sun and a planet as seen from Earth. In the case of Mercury or Venus the largest separation is referred to as greatest elongation either east or west of the Sun.

**Ephemeris:** A table of computed positions of celestial objects for a given date.

**Ingress:** The beginning of a transit event.

**Irradiation:** A deceptive trick of the eye and brain whereby bright objects set against a dark background can appear larger than they truly are. Using certain filters helps to reduce the effect.

**Latitude:** A position 0° to +90° north or 0° to –90° south of the equator of a celestial body.

**Libration:** A term most often associated with the Moon, concerning slight oscillations during its accelerated and decelerated orbit around the Earth. This effect means we can view more of the lunar surface to the east, west (libration in longitude), north and south (libration in latitude) than a theoretical 50 per cent.

**Limb:** The apparent edge of the Sun, Moon, planet or other celestial body with a detectable disc.

**Magnitude:** Also known as visual stellar magnitude, it is the measure of the apparent brightness of a celestial object.

**Meridian:** A great imaginary circle that runs from north through the observer's zenith to south, then through nadir directly below the observer back to the originating northern point.

**Nadir:** The point directly below the observer or 90° vertically from the horizon.

**Oblateness:** The flattening of a spheroid, such as can be seen in Jupiter. It is calculated by dividing the difference between the equatorial and polar diameters by the equatorial diameter.

**Occultation:** The hiding of one object in the sky by another; when the Moon passes in front of a planet or a star is one example.

**Opposition:** The point where a planet or other body is opposite the Sun in our sky.

**Perigee:** The point at which a body in orbit around the Earth, such as the Moon, is nearest the Earth during its orbit.

**Perihelion:** Point at which a planet or other Sun-orbiting body is nearest the Sun.

**Perturbations:** An oscillation or irregularity in the orbit or motion of a celestial body, caused by the gravitational influence of other celestial bodies.

**Phase:** The amount of illumination of surface area of a planet or the Moon from reflected sunlight as seen from Earth.

**Planetesimal:** A small body, perhaps a few millimetres to over 1 kilometre in size, that has accreted from the solar nebula.

**Quadrature:** The position of a planet or the Moon when it is at 90° with respect to the Earth and the Sun.

**Revolution:** The motion in which a body moves around another body, such as the planets around the Sun or the Moon around the Earth.

**Right Ascension (RA):** The term given to the east-to-west motion of the celestial sphere. Since Earth completes one axial rotation through 360° every 24 hours, the sky can subsequently be divided into hours of arc. Hence one hour of arc represents 15°. For refined positional accuracy each hour is further subdivided into minutes and seconds of arc.

**Rotation:** The motion of a body when turning on its axis.

**Sidereal period:** The time taken for one revolution of a celestial body around another with respect to the stars.

**Terminator:** The sunset or sunrise line on the Moon or a planet. The boundary between night and day.

**Transit:** The passing of one celestial body in front of another or across a line or a circle, for example Mercury passing across the face of the Sun, a Jovian satellite across the face of Jupiter, or a star across the meridian.

**Zenith:** The point in the sky directly above the observer or 90º vertically from the horizon.

# References and further reading


Aguirre, EL 2001, 'Astro imaging with digital cameras', *Sky & Telescope Magazine*, vol. 102, no. 2, p. 128.

Bakich, M 2000, *The Cambridge planetary handbook*, Cambridge University Press, Cambridge.

Batson, R & Greely, R 1997, *The NASA atlas of the solar system*, Cambridge University Press, Cambridge.

Boyce, JM 2002, *The Smithsonian book of Mars*, Smithsonian Institution Press, Washington DC.

Dantowitz, R 1998, 'Sharper images through video', *Sky & Telescope Magazine*, vol. 96, no. 2, p. 48.

Dobbins, T, Douglas, E & Massey, S 2000, *Video Astronomy*, Sky Publishing, Cambridge.

Golombek, M & Raeburn, P 1998, *Mars (uncovering the secrets of the red planet)*, National Geographic Society, Washington DC.

Goodwin, S & Gribbin, J 1997, *Origins*, Constable & Co., London.

Greely, R, Malin, MC & Murray, B 1981, *Earthlike planets* WH Freeman & Co., San Francisco.

Grinspoon, DH 1997, *Venus revealed*, Addison-Wesley Publishing Co Inc., New York.

Hoskin, M 1997, *Cambridge illustrated history of astronomy*, Cambridge University Press, Cambridge.

Johnson, TV, McFadden, L & Weissman, PR 1999, *Encyclopedia of the solar system*, Academic Press, California.

Kaufmann, WJ 1979, *Planets and moons*, WH Freeman & Co., San Francisco.

Massey, S 2000, 'Return of the giants', *Sky & Space Magazine*, vol. 13, no. 5, issue 62, p. 54.

Moore, P 1998 *On Mars*, Cassell, London.

Moore, P (ed.) *2002, Philip's astronomy encyclopedia*, Octopus Publishing Group, London.

Rükl, A 1996, *Atlas of the moon*, Kalmbach Publishing Co., Wisconsin.

Schorn, RA 1998, *Planetary astronomy*, Texas A&M University Press, College Station.

Troiani, DM 2001, 'A grand return of Mars', *Sky & Telescope Magazine*, vol. 101, no. 5, p. 201.

# Index

Page numbers in **bold** refer to illustrations